Electric Machines and Drives

A First Course

Electric Machines and Drives

A First Course

NED MOHAN
Department of Electrical and Computer Engineering
University of Minnesota, Minneapolis
USA

WILEY
John Wiley & Sons, Inc.

VP & PUBLISHER:	Don Fowley
EDITOR:	Dan Sayre
PROJECT EDITOR:	Nithyanand Rao
EDITORIAL ASSISTANT:	Charlotte Cerf
MARKETING MANAGER:	Christopher Ruel
MARKETING ASSISTANT:	Ashley Tomeck
DESIGNER:	James O'Shea
SENIOR PRODUCTION MANAGER:	Janis Soo
SENIOR PRODUCTION EDITOR:	Joyce Poh

PSpice is a registered trademark of the OrCAD Corporation.
SIMULINK is a registerd trademark of The Mathworks, Inc.

This book was set in 10/12 TimesNewRoman by MPS Limited, a Macmillan Company, Chennai, India and printed and bound by Hamilton Printing. The cover was printed by Hamilton Printing.

This book is printed on acid free paper.

Founded in 1807, John Wiley & Sons, Inc. has been a valued source of knowledge and understanding for more than 200 years, helping people around the world meet their needs and fulfill their aspirations. Our company is built on a foundation of principles that include responsibility to the communities we serve and where we live and work. In 2008, we launched a Corporate Citizenship Initiative, a global effort to address the environmental, social, economic, and ethical challenges we face in our business. Among the issues we are addressing are carbon impact, paper specifications and procurement, ethical conduct within our business and among our vendors, and community and charitable support. For more information, please visit our website: www.wiley.com/go/citizenship.

Evaluation copies are provided to qualified academics and professionals for review purposes only, for use in their courses during the next academic year. These copies are licensed and may not be sold or transferred to a third party. Upon completion of the review period, please return the evaluation copy to Wiley. Return instructions and a free of charge return mailing label are available at www.wiley.com/go/returnlabel. If you have chosen to adopt this textbook for use in your course, please accept this book as your complimentary desk copy. Outside of the United States, please contact your local sales representative.

Library of Congress Cataloging-in-Publication Data

Mohan, Ned.
    Electric machines and drives : a first course / Ned Mohan.
      p. cm.
    Includes bibliographical references and index.
    ISBN 978-1-118-07481-7 (hardback : acid free paper)
      1. Electric machinery. 2. Electric driving. I. Title.
TK2000.M57 2012
621.31′042—dc23

                        2011043892

Printed in the United States of America
10 9 8 7 6 5 4 3 2 1

CONTENTS

PREFACE

Role of Electric Machines and Drives in Sustainable Electric Energy Systems

Sustainable electric energy systems require that we utilize renewable sources for generating electricity and use it as efficiently as possible. Towards this goal, electric machines and drives are required for harnessing wind energy, for example. Nearly one-half to two-thirds of the electric energy we use is consumed by motor-driven systems. In most such applications, it is possible to obtain a much higher system-efficiency by appropriately varying the rotational speed based on the operating conditions. Another emerging application of variable-speed drives is in electric vehicles and hybrid-electric vehicles.

This textbook explains the basic principles on which electric machines operate and how their speed can be controlled efficiently.

A New Approach

This textbook is intended for a first course on the subject of electric machines and drives where no prior exposure to this subject is assumed. To do so in a single-semester course, a physics-based approach is used that not only leads to thoroughly understanding the basic principles on which electric machines operate, but also shows how they ought to be controlled for maximum efficiency. Moreover, electric machines are covered as a part of electric-drive systems, including power electronic converters and control, hence allowing relevant and interesting applications in wind-turbines and electric vehicles, for example, to be discussed.

This textbook describes systems under steady state operating conditions. However, the uniqueness of the approach used here seamlessly allows the discussion to be continued for analyzing and controlling of systems under dynamic conditions that are discussed in graduate-level courses.

1

INTRODUCTION TO ELECTRIC DRIVE SYSTEMS

Spurred by advances in power electronics, adjustable-speed electric drives now offer great opportunities in a plethora of applications: pumps and compressors to save energy, precision motion control in automated factories, and wind-electric systems to generate electricity, to name a few. A recent example is the commercialization of hybrid-electric vehicles [1]. Figure 1.1 shows the photograph of a hybrid arrangement in which the outputs of the internal-combustion (IC) engine and the electric drive are mechanically added in parallel to drive the wheels. Compared to vehicles powered solely by gasoline, these hybrids reduce fuel consumption by more than 50 percent and emit far fewer pollutants.

1.1 HISTORY

Electric machines have now been in existence for over a century. All of us are familiar with the basic function of electric motors: to drive mechanical loads by converting electrical energy. In the absence of any control, electric motors operate at essentially a constant speed. For example, when the compressor-motor in a refrigerator turns on, it runs at a constant speed.

FIGURE 1.1 Photograph of a hybrid-electric vehicle.

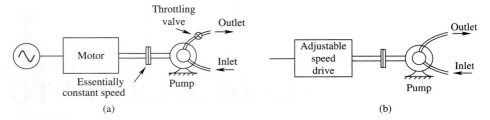

FIGURE 1.2   Traditional and ASD-based flow control systems.

Traditionally, motors were operated uncontrolled, running at constant speeds, even in applications where efficient control over their speed could be very advantageous. For example, consider the process industry (like oil refineries and chemical factories) where the flow rates of gases and fluids often need to be controlled. As Figure 1.2a illustrates, in a pump driven at a constant speed, a throttling valve controls the flow rate. Mechanisms such as throttling valves are generally more complicated to implement in automated processes and waste large amounts of energy. In the process industry today, electronically controlled adjustable-speed drives (ASDs), shown in Figure 1.2b, control the pump speed to match the flow requirement. Systems with adjustable-speed drives are much easier to automate, and offer much higher energy efficiency and lower maintenance than the traditional systems with throttling valves.

These improvements are not limited to the process industry. Electric drives for speed and position control are increasingly being used in a variety of manufacturing, heating, ventilating and air conditioning (HVAC), and transportation systems.

1.2   WHAT IS AN ELECTRIC-MOTOR DRIVE?

Figure 1.3 shows the block diagram of an electric-motor drive, or for short, an electric drive. In response to an input command, electric drives efficiently control the speed and/ or the position of the mechanical load, thus eliminating the need for a throttling valve like the one shown in Figure 1.2a. The controller, by comparing the input command for speed and/or position with the actual values measured through sensors, provides appropriate control signals to the power-processing unit (PPU) consisting of power semiconductor devices.

As Figure 1.3 shows, the power-processing unit gets its power from the utility source with single-phase or three-phase sinusoidal voltages of a fixed frequency and constant amplitude.

The power-processing unit, in response to the control inputs, efficiently converts these fixed-form input voltages into an output of the appropriate form (in frequency, amplitude, and the number of phases) that is optimally suited for operating the motor. The input command to the electric drive in Figure 1.3 may come from a process computer, which considers the objectives of the overall process and issues a command to control the mechanical load. However, in general-purpose applications, electric drives operate in an open-loop manner without any feedback.

Throughout this text, we will use the term *electric-motor drive* (*motor drive* or *drive* for short) to imply the combination of blocks in the box drawn by dotted lines in Figure 1.3. We will examine all of these blocks in subsequent chapters.

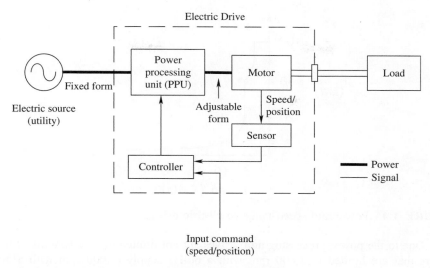

FIGURE 1.3 Block diagram of an electric drive system.

1.3 FACTORS RESPONSIBLE FOR THE GROWTH OF ELECTRIC DRIVES

Technical Advancements. Controllers used in electric drives (see Figure 1.3) have benefited from revolutionary advances in microelectronic methods, which have resulted in powerful linear integrated circuits and digital signal processors [2]. These advances in semiconductor fabrication technology have also made it possible to significantly improve voltage- and current-handling capabilities, as well as the switching speeds of power semiconductor devices, which make up the power-processing unit of Figure 1.3.

Market Needs. The world market of adjustable-speed drives was estimated as a $20 billion industry in 1997. This market is growing at a healthy rate [3] as users discover the benefits of operating motors at variable speeds. These benefits include improved process control, reduction in energy usage, and less maintenance.

The world market for electric drives would be significantly impacted by large-scale opportunities for harnessing wind energy. There is also a large potential for applications in the developing world, where the growth rates are the highest.

Applications of electric drives in the United States are of particular importance. The per-capita energy consumption in the United States is almost twice that in Europe, but the electric drive market in 1997 was less than one-half. This deficit, due to a relatively low cost of energy in the United States, represents a tremendous opportunity for application of electric drives.

1.4 TYPICAL APPLICATIONS OF ELECTRIC DRIVES

Electric drives are increasingly being used in most sectors of the economy. Figure 1.4 shows that electric drives cover an extremely large range of power and speed — up to 100 MW in power and up to 80,000 rpm in speed.

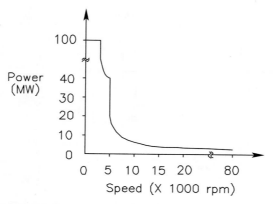

FIGURE 1.4 Power and speed range of electric drives.

Due to the power-processing unit, drives are not limited in speeds, unlike line-fed motors that are limited to 3,600 rpm with a 60-Hz supply (3,000 rpm with a 50-Hz supply). A large majority of applications of drives are in a low-to-medium power range, from a fractional kW to several hundred kW. Some of these application areas are listed below:

- Process Industry: agitators, pumps, fans, and compressors
- Machining: planers, winches, calendars, chippers, drill presses, sanders, saws, extruders, feeders, grinders, mills, and presses
- Heating, Ventilating and Air Conditioning: blowers, fans, and compressors
- Paper and Steel Industry: hoists, and rollers
- Transportation: elevators, trains, and automobiles
- Textile: looms
- Packaging: shears
- Food: conveyors, and fans
- Agriculture: dryer fans, blowers, and conveyors
- Oil, Gas, and Mining: compressors, pumps, cranes, and shovels
- Residential: heat pumps, air conditioners, freezers, appliances, and washing machines

In the following sections, we will look at a few important applications of electric drives in energy conservation, wind-electric generation, and electric transportation.

1.4.1 Role of Drives in Energy Conservation [4]

It is perhaps not obvious how electric drives can reduce energy consumption in many applications. Electric costs are expected to continue their upward trend, which makes it possible to justify the initial investment in replacing constant-speed motors with adjustable-speed electric drives, solely on the basis of reducing energy expenditure (see Chapter 15). The environmental impact of energy conservation, in reducing global warming and acid rain, is also of vital importance [5].

To arrive at an estimate of the potential role of electric drives in energy conservation, consider that the motor-driven systems in the United States are responsible for

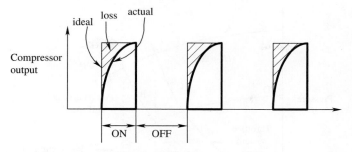

FIGURE 1.5 Heat pump operation with line-fed motors.

over 57% of all electric power generated and 20% of all the energy consumed. The United States Department of Energy estimates that if constant-speed, line-fed motors in pump and compressor systems were to be replaced by adjustable-speed drives, the energy efficiency would improve by as much as 20%. This improved energy efficiency amounts to huge potential savings (see homework problem 1.1). In fact, the potential yearly energy savings would be approximately equal to the annual electricity use in the state of New York. Some energy-conservation applications are described as follows.

1.4.1.1 Heat Pumps and Air Conditioners [6]

Conventional air conditioners cool buildings by extracting energy from inside the building and transferring it to the atmosphere outside. Heat pumps, in addition to the air-conditioning mode, can also heat buildings in winter by extracting energy from outside and transferring it inside. The use of heat pumps for heating and cooling is on the rise; they are now employed in roughly one out of every three new homes constructed in the United States (Figure 1.5).

In conventional systems, the building temperature is controlled by on/off cycling of the compressor motor by comparing the building temperature with the thermostat setting. After being off, when the compressor motor turns on, the compressor output builds up slowly (due to refrigerant migration during the off period) while the motor immediately begins to draw full power. This cyclic loss (every time the motor turns on) between the ideal and the actual values of the compressor output, as shown in Figure 1.6, can be eliminated by running the compressor continuously at a speed at which its output matches the thermal load of the building. Compared to conventional systems, compressors driven by adjustable speed drives reduce power consumption by as much as 30 percent.

1.4.1.2 Pumps, Blowers, and Fans

To understand the savings in energy consumption, let us compare the two systems shown in Figure 1.2. In Figure 1.6, curve A shows the full-speed pump characteristic; that is, the pressure (or head) generated by a pump, driven at its full speed, as a function of flow rate. With the throttling valve fully open, curve B shows the unthrottled system characteristic, that is, the pressure required as a function of flow rate, to circulate fluid or gas by overcoming the static potential (if any) and friction. The full flow rate Q_1 is given by the intersection of the unthrottled system curve B with the pump curve A. Now consider that a reduced flow rate Q_2 is desired, which requires a pressure H_2 as seen from the unthrottled system curve B. Below, we will consider two ways of achieving this reduced flow rate.

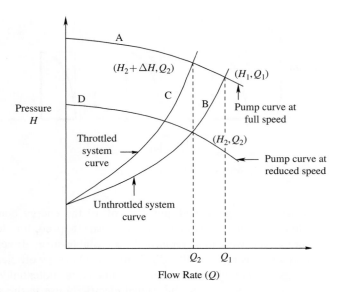

FIGURE 1.6   Typical pump and system curve.

With a constant-speed motor as in Figure 1.2a, the throttling valve is partially closed, which requires additional pressure to be overcome by the pump, such that the throttled system curve C intersects with the full-speed pump curve A at the flow rate Q_2. The power loss in the throttling valve is proportional to Q_2 times ΔH. Due to this power loss, the reduction in energy efficiency will depend on the reduced flow-rate intervals, compared to the duration of unthrottled operation.

The power loss across the throttling valve can be eliminated by means of an adjustable-speed drive. The pump speed is reduced such that the reduced-speed pump curve D in Figure 1.6 intersects with the unthrottled system curve B at the desired flow rate Q_2.

Similarly, in blower applications, the power consumption can be substantially lowered, as plotted in Figure 1.7, by reducing the blower speed by means of an adjustable speed drive to decrease flow rates, rather than using outlet dampers or inlet vanes. The percentage reduction in power consumption depends on the flow-rate profile (see homework problem 1.5).

Electric drives can be beneficially used in almost all pumps, compressors, and blowers employed in air-handling systems, process industries, and the generating plants of electric utilities. There are many documented examples where energy savings alone have paid for the cost of conversion (from line-fed motors to electric-drive systems) within six months of operation. Of course, this advantage of electric drives is made possible by the ability to control motor speeds in an energy-efficient manner, as discussed in the subsequent chapters.

1.4.2   Harnessing Wind Energy

Electric drives also play a significant role in power generation from renewable energy sources, such as wind and small hydro. The block diagram for a wind-electric system is shown in Figure 1.8, where the variable-frequency ac produced by the wind-turbine−driven

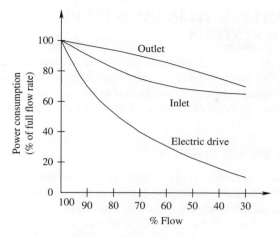

FIGURE 1.7 Power consumption in a blower.

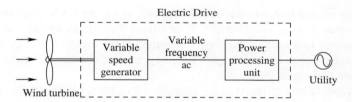

FIGURE 1.8 Electric drive for wind generators.

generator is interfaced with the utility system through a power-processing unit. By letting the turbine speed vary with the wind speed, it is possible to recover a higher amount of energy compared to systems where the turbine essentially rotates at a constant speed due to the generator output being directly connected to the utility grid [7]. Harnessing of wind energy is turning out to be a major application of electric drives and this sector is expected to grow rapidly.

1.4.3 Electric Transportation

Electric transportation is widely used in many countries. Experiments with magnetically-levitated trains are being conducted in Japan and Germany. High-speed electric trains are also presently being evaluated in the United States for mass transportation in the northeastern and southwestern corridors.

Another important application of electric drives is in electric vehicles and hybrid-electric vehicles. The main virtue of electric vehicles (especially to large metropolitan areas) is that they emit no pollutants. However, electric vehicles must wait for suitable batteries, fuel cells, or flywheels to be developed before the average motorist accepts them. On the other hand, hybrid-electric vehicles are already commercialized [1].

There are many new applications of electric drives in conventional automobiles. Also, there is an ongoing attempt to replace hydraulic drives with electric drives in airplanes and ships.

1.5 THE MULTI-DISCIPLINARY NATURE
OF DRIVE SYSTEMS

The block diagram of Figure 1.3 points to various fields that are essential to electric drives: electric machine theory, power electronics, analog and digital control theory, real-time application of digital controllers, mechanical system modeling, and interaction with electric-power systems. A brief description of each of the fields is provided in the following subsections.

1.5.1 Theory of Electric Machines

For achieving the desired motion, it is necessary to control electric motors appropriately. This requires a thorough understanding of the operating principles of various commonly used motors such as dc, synchronous, induction and stepper motors. The emphasis in an electric drives course needs to be different from that in traditional electric machines courses, which are oriented towards design and application of line-fed machines.

1.5.2 Power Electronics

The discipline related to the power-processing unit in Figure 1.3 is often referred to as power electronics. Voltages and currents from a fixed form (in frequency and magnitude) must be converted to the adjustable form best suited to the motor. It is important that the conversion take place at a high energy efficiency, which is realized by operating power semiconductor devices as switches.

Today, power processing is being simplified by means of "Smart Power" devices, where power semiconductor switches are integrated with their protection and gate-drive circuits into a single module. Thus, the logic-level signals (such as those supplied by a digital signal processor) can directly control high-power switches in the PPU. Such power-integrated modules are available with voltage-handling capability approaching 4 kilovolts and current-handling capability in excess of 1,000 amperes. Paralleling such modules allows even higher current-handling capabilities.

The progress in this field has made a dramatic impact on power-processing units by reducing their size and weight, while substantially increasing the number of functions that can be performed [3].

1.5.3 Control Theory

In the majority of applications, the speed and position of drives need not be controlled precisely. However, there are an increasing number of applications, for example in robotics for automated factories, where accurate control of torque, speed, and position are crucial. Such control is accomplished by feeding back the measured quantities, and by comparing them with their desired values, in order to achieve a fast and accurate control.

In most motion-control applications, it is sufficient to use a simple proportional-integral (PI) control as discussed in Chapter 8. The task of designing and analyzing PI-type controllers is made easy due to the availability of powerful simulation tools such as PSpice.

1.5.4 Real-Time Control Using DSPs

All modern electric drives use microprocessors and digital signal processors (DSPs) for flexibility of control, fault diagnosis, and communication with the host computer and with other process computers. Use of 8-bit microprocessors is being replaced by 16-bit and even 32-bit microprocessors. Digital signal processors are used for real-time control in applications that demand high performance or where a slight gain in the system efficiency more than pays for the additional cost of a sophisticated control.

1.5.5 Mechanical System Modeling

Specifications of electric drives depend on the torque and speed requirements of the mechanical loads. Therefore, it is often necessary to model mechanical loads. Rather than considering the mechanical load and the electric drive as two separate subsystems, it is preferable to consider them together in the design process. This design philosophy is at the heart of Mechatronics.

1.5.6 Sensors

As shown in the block diagram of electric drives in Figure 1.3, voltage, current, speed, and position measurements may be required. For thermal protection, the temperature needs to be sensed.

1.5.7 Interactions of Drives with the Utility Grid

Unlike line-fed electric motors, electric motors in drives are supplied through a power electronic interface (see Figure 1.3). Therefore, unless corrective action is taken, electric drives draw currents from the utility that are distorted (nonsinusoidal) in wave shape. This distortion in line currents interferes with the utility system, degrading its power quality by distorting the voltages. Available technical solutions make the drive interaction with the utility harmonious, even more so than line-fed motors. The sensitivity of drives to power-system disturbances such as sags, swells, and transient overvoltages should also be considered. Again, solutions are available to reduce or eliminate the effects of these disturbances.

1.6 STRUCTURE OF THE TEXTBOOK

Chapter 1 has introduced the roles and applications of electric drives. Chapter 2 deals with the modeling of mechanical systems coupled to electric drives, as well as how to determine drive specifications for various types of loads. Chapter 3 reviews linear electric circuits. An introduction to power-processing units is presented in Chapter 4.

Magnetic circuits, including transformers, are discussed in Chapter 5. Chapter 6 explains the basic principles of electromagnetic energy conversion.

Chapter 7 describes dc-motor drives. Although the share of dc-motor drives in new applications is declining, their use is still widespread. Another reason for studying dc-motor drives is that ac-motor drives are controlled to emulate their performance. The feedback-controller design for drives (using dc drives as an example) is presented in Chapter 8.

As a background to the discussion of ac motor drives, the rotating fields in ac machines are described in Chapter 9 by means of space vectors. Using the space-vector theory, the sinusoidal waveform PMAC motor drives are discussed in Chapter 10. Chapter 11 introduces induction motors and focuses on their basic principles of operation in steady state. A concise but comprehensive discussion of controlling speed with induction-motor drives is provided in Chapter 12. The reluctance drives, including stepper-motors and switched-reluctance drives, are explained in Chapter 13. Loss considerations and various techniques to improve energy efficiency in drives are discussed in Chapter 14.

SUMMARY/REVIEW QUESTIONS

1. What is an electric drive? Draw the block diagram and explain the roles of its various components.
2. What has been the traditional approach to controlling flow rate in the process industry? What are the major disadvantages that can be overcome by using adjustable-speed drives?
3. What are the factors responsible for the growth of the adjustable-speed drive market?
4. How does an air conditioner work?
5. How does a heat pump work?
6. How do ASDs save energy in air-conditioning and heat-pump systems?
7. What is the role of ASDs in industrial systems?
8. There are proposals to store energy in flywheels for load leveling in utility systems. During the off-peak period for energy demand at night, these flywheels are charged to high speeds. At peak periods during the day, this energy is supplied back to the utility. How would ASDs play a role in this scheme?
9. What is the role of electric drives in electric transportation systems of various types?
10. List a few specific examples from the applications mentioned in section 1.4 that you are personally familiar with.
11. What are the different disciplines that make up the study and design of electric-drive systems?

REFERENCES

1. V. Wouk et al., E.V. Watch, *IEEE Spectrum* (March 1998): 22–23.
2. N. Mohan, T. Undeland, and W. Robbins, *Power Electronics: Converters, Applications, and Design*, 2nd ed. (New York: John Wiley & Sons, 1995).
3. P. Thogersen and F. Blaabjerg, "Adjustable-Speed Drives in the Next Decade: The Next Steps in Industry and Academia," Proceedings of the PCIM Conference, Nuremberg, Germany, June 6–8, 2000.
4. N. Mohan, *Techniques for Energy Conservation in AC Motor Driven Systems*, Electric Power Research Institute Final Report EM-2037, Project 1201-1213, September 1981.
5. Y. Kaya, "Response Strategies for Global Warming and the Role of Power Technologies," Proceedings of the IPEC, Tokyo, Japan, April 3–5, 2000, pp. 1–3.
6. N. Mohan and J. Ramsey, *Comparative Study of Adjustable-Speed Drives for Heat Pumps*, Electric Power Research Institute Final Report EM-4704, Project 2033-4, August 1986.
7. F. Blaabjerg and N. Mohan, "Wind Power," *Encyclopedia of Electrical and Electronics Engineering*, edited by John G. Webster (New York: John Wiley & Sons, 1998).
8. D.M. Ionel, "High-efficiency variable-speed electric motor drive technologies for energy savings in the US residential sector," 12th International Conference on Optimization of Electrical and Electronic Equipment, OPTIM 2010, Brasov, Romania, ISSN: 1842-0133.
9. Kara Clark, Nicholas W. Miller, Juan J. Sanchez-Gasca, *Modeling of GE Wind Turbine-Generators for Grid Studies*, GE Energy Report, Version 4.4, September 9, 2009.

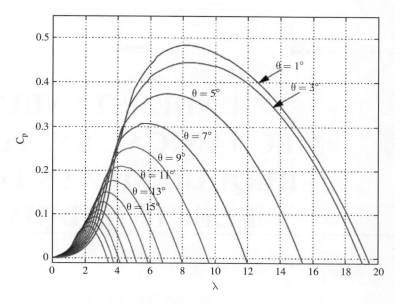

FIGURE P1.5 Plot of C_p as a function of λ [9].

PROBLEMS

1.1 A U.S. Department of Energy report estimates that over 100 billion kWh/year can be saved in the United States by various energy-conservation techniques applied to the pump-driven systems. Calculate (a) how many 1000-MW generating plants running constantly supply this wasted energy and (b) the annual savings in dollars if the cost of electricity is 0.10 $/kWh.

1.2 Visit your local machine-tool shop and make a list of various electric drive types, applications, and speed/torque ranges.

1.3 Repeat Problem 1.2 for automobiles.

1.4 Repeat Problem 1.2 for household appliances [8].

1.5 In wind-turbines, the ratio (P_{shaft}/P_{wind}) of the power available at the shaft to the power in the wind is called the Coefficient of Performance, C_p, which is a unit-less quantity. For informational purposes, the plot of this coefficient, as a function of λ is shown in Figure P1-5 [9] for various values of the blades pitch-angle θ, where λ is a constant times the ratio of the blade-tip speed and the wind speed.

The rated power is produced at the wind speed of 12 m/s where the rotational speed of the blades is 20 rpm. The cut-in wind speed is 4 m/s. Calculate the range over which the blade speed should be varied, between the cut-in and the rated wind speeds, to harness the maximum power from the wind. In this range of wind speeds, the blades' pitch-angle θ is kept at nearly zero. Note: this simple problem shows the benefit of varying the speed of wind-turbines, by means of a variable-speed drive, for maximizing the harnessed energy.

2

UNDERSTANDING MECHANICAL SYSTEM REQUIREMENTS FOR ELECTRIC DRIVES

2.1 INTRODUCTION

Electric drives must satisfy the requirements of torque and speed imposed by mechanical loads connected to them. The load in Figure 2.1, for example, may require a trapezoidal profile for the angular speed, as a function of time. In this chapter, we will briefly review the basic principles of mechanics for understanding the requirements imposed by mechanical systems on electric drives. This understanding is necessary for selecting an appropriate electric drive for a given application.

2.2 SYSTEMS WITH LINEAR MOTION

In Figure 2.2a, a load of a constant mass M is acted upon by an external force f_e that causes it to move in the linear direction x at a speed $u = dx/dt$.

This movement is opposed by the load, represented by a force f_L. The linear momentum associated with the mass is defined as M times u. As shown in Figure 2.2b, in

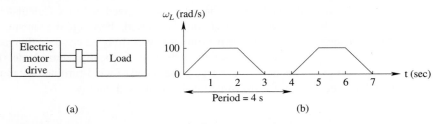

FIGURE 2.1 (a) Electric drive system; (b) Example of load-speed profile requirement.

12

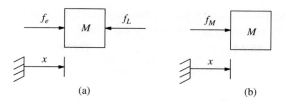

FIGURE 2.2 Motion of a mass M due to action of forces.

accordance with the Newton's Law of Motion, the net force $f_M (= f_e - f_L)$ equals the rate of change of momentum, which causes the mass to accelerate:

$$f_M = \frac{d}{dt}(Mu) = M\frac{du}{dt} = Ma \tag{2.1}$$

where a is the acceleration in m/s^2, which from Equation 2.1 is

$$a = \frac{du}{dt} = \frac{f_M}{M} \tag{2.2}$$

In MKS units, a net force of 1 Newton (or 1 N), acting on a constant mass of 1 kg results in an acceleration of 1 m/s^2. Integrating the acceleration with respect to time, we can calculate the speed as

$$u(t) = u(0) + \int_0^t a(\tau) \cdot d\tau \tag{2.3}$$

and, integrating the speed with respect to time, we can calculate the position as

$$x(t) = x(0) + \int_0^t u(\tau) \cdot d\tau \tag{2.4}$$

where τ is a variable of integration.

The differential work dW done by the mechanism supplying the force f_e is

$$dW_e = f_e\, dx \tag{2.5}$$

Power is the time-rate at which the work is done. Therefore, differentiating both sides of Equation 2.5 with respect to time t, and assuming that the force f_e remains constant, the power supplied by the mechanism exerting the force f_e is

$$p_e(t) = \frac{dW_e}{dt} = f_e\frac{dx}{dt} = f_e u \tag{2.6}$$

It takes a finite amount of energy to bring a mass to a speed from rest. Therefore, a moving mass has stored kinetic energy that can be recovered. Note that in the system of Figure 2.2, the net force $f_M (= f_e - f_L)$ is responsible for accelerating the mass. Therefore, assuming that f_M remains constant, the net power $p_M(t)$ going into accelerating the mass can be calculated by replacing f_e in Equation 2.6 with f_M:

$$p_M(t) = \frac{dW_M}{dt} = f_M \frac{dx}{dt} = f_M u \tag{2.7}$$

From Equation 2.1, substituting f_M as $M \frac{du}{dt}$,

$$p_M(t) = Mu \frac{du}{dt} \tag{2.8}$$

The energy input, which is stored as kinetic energy in the moving mass, can be calculated by integrating both sides of Equation 2.8 with respect to time. Assuming the initial speed u to be zero at time $t = 0$, the stored kinetic energy in the mass M can be calculated as

$$W_M = \int_0^t p_M(\tau)d\tau = M \int_0^t u \frac{du}{d\tau} d\tau = M \int_0^u u\, du = \tfrac{1}{2} Mu^2 \tag{2.9}$$

where τ is a variable of integration.

2.3 ROTATING SYSTEMS

Most electric motors are of rotating type. Consider a lever, pivoted and free to move as shown in Figure 2.3a. When an external force f is applied in a *perpendicular* direction at a radius r from the pivot, then the torque acting on the lever is

$$\begin{matrix} T \ = \ f & r \\ \text{[Nm]} \quad \text{[N]} & \text{[m]} \end{matrix} \tag{2.10}$$

which acts in a counter-clockwise direction, considered here to be positive.

Example 2.1

In Figure 2.3a, a mass M is hung from the tip of the lever. Calculate the holding torque required to keep the lever from turning, as a function of angle θ in the range of 0 to 90 degrees. Assume that $M = 0.5$ kg and $r = 0.3$ m.

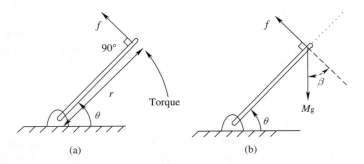

(a)   (b)

FIGURE 2.3 (a) Pivoted Lever; (b) Holding torque for the lever.

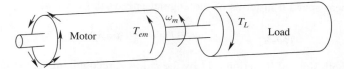

FIGURE 2.4 Torque in an electric motor.

Solution The gravitational force on the mass is shown in Figure 2.3b. For the lever to be stationary, the net force perpendicular to the lever must be zero; that is, $f = M \, g \cos \beta$ where $g = 9.8 \text{ m/s}^2$ is the gravitational acceleration. Note in Figure 2.3b that $\beta = \theta$. The holding torque T_h must be $T_h = fr = Mgr \cos \theta$. Substituting the numerical values,

$$T_h = 0.5 \times 9.8 \times 0.3 \times \cos \theta = 1.47 \cos \theta \ \text{Nm}$$

In electric machines, the various forces shown by arrows in Figure 2.4 are produced due to electromagnetic interactions. The definition of torque in Equation 2.10 correctly describes the resulting electromagnetic torque T_{em} that causes the rotation of the motor and the mechanical load connected to it by a shaft.

In a rotational system, the angular acceleration due to the net torque acting on it is determined by its moment-of-inertia J. The example below shows how to calculate the moment-of-inertia J of a rotating solid cylindrical mass.

Example 2.2

    a. Calculate the moment-of-inertia J of a solid cylinder that is free to rotate about its axis, as shown in Figure 2.5a, in terms of its mass M and the radius r_1.

    b. Given that a solid steel cylinder has radius $r_1 = 6$ cm, length $\ell = 18$ cm, and the material density $\rho = 7.85 \times 10^3 \text{ kg/m}^3$, calculate its moment-of-inertia J.

Solution (a) From Newton's Law of Motion, in Figure 2.5a, to accelerate a differential mass dM at a radius r, the net differential force df required in a perpendicular (tangential) direction, from Equation 2.1, is

$$(dM)\left(\frac{du}{dt}\right) = df \tag{2.11}$$

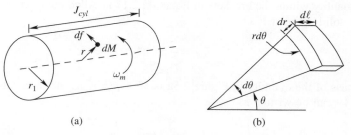

(a)   (b)

FIGURE 2.5 Calculation of the inertia, J_{cyl}, of a solid cylinder.

where the linear speed u in terms of the angular speed ω_m (in rad/s) is

$$u = r\,\omega_m \qquad (2.12)$$

Multiplying both sides of Equation 2.11 by the radius r, recognizing that $(r\,df)$ equals the net differential torque dT and using Equation 2.12,

$$r^2\,dM\,\frac{d}{dt}\omega_m = dT \qquad (2.13)$$

The same angular acceleration $\frac{d}{dt}\omega_m$ is experienced by all elements of the cylinder. With the help of Figure 2.5b, the differential mass dM in Equation 2.13 can be expressed as

$$dM = \rho\,\underbrace{r d\theta}_{arc}\,\underbrace{dr}_{height}\,\underbrace{d\ell}_{length} \qquad (2.14)$$

where ρ is the material density in kg/m^3. Substituting dM from Equation 2.14 into Equation 2.13,

$$\rho\,(r^3 dr\,d\theta\,d\ell)\frac{d}{dt}\omega_m = dT \qquad (2.15)$$

The net torque acting on the cylinder can be obtained by integrating over all differential elements in terms of r, θ, and ℓ as

$$\rho\left(\int_0^{r_1} r^3 dr \int_0^{2\pi} d\theta \int_0^{\ell} d\ell\right)\frac{d}{dt}\omega_m = T. \qquad (2.16)$$

Carrying out the triple integration yields

$$\underbrace{\left(\frac{\pi}{2}\rho\,\ell\,r_1^4\right)}_{J_{cyl}}\frac{d}{dt}\omega_m = T \qquad (2.17)$$

or

$$J_{cyl}\frac{d\omega_m}{dt} = T \qquad (2.18)$$

where the quantity within the brackets in Equation 2.17 is called the moment-of-inertia J, which for a solid cylinder is

$$J_{cyl} = \frac{\pi}{2}\rho\,\ell\,r_1^4. \qquad (2.19)$$

Since the mass of the cylinder in Figure 2.5a is $M = \rho(\pi\,r_1^2)\ell$, the moment-of-inertia in Equation 2.19 can be written as

$$J_{cyl} = \frac{1}{2}M\,r_1^2 \qquad (2.20)$$

(b) Substituting $r_1 = 6$ cm, length $\ell = 18$ cm, and $\rho = 7.85 \times 10^3$ kg/m³ in Equation 2.19, the moment-of-inertia J_{cyl} of the cylinder in Figure 2.5a is

$$J_{cyl} = \frac{\pi}{2} \times 7.85 \times 10^3 \times 0.18 \times (0.06)^4 = 0.029 \text{ kg} \cdot \text{m}^2$$

The net torque T_J acting on the rotating body of inertia J causes it to accelerate. Similar to systems with linear motion where $f_M = M\,a$, Newton's Law in rotational systems becomes

$$T_J = J\,\alpha \tag{2.21}$$

where the angular acceleration $\alpha (= d\omega/dt)$ in rad/s² is

$$\alpha = \frac{d\omega_m}{dt} = \frac{T_J}{J} \tag{2.22}$$

which is similar to Equation 2.18 in the previous example. In MKS units, a torque of 1 Nm, acting on an inertia of 1 kg·m² results in an angular acceleration of 1 rad/s².

In systems such as the one shown in Figure 2.6a, the motor produces an electro-magnetic torque, T_{em}. The bearing friction and wind resistance (drag) can be combined with the load torque T_L opposing the rotation. In most systems, we can assume that the rotating part of the motor with inertia J_M is rigidly coupled (without flexing) to the load inertia J_L. The net torque, the difference between the electromagnetic torque developed by the motor and the load torque opposing it, causes the combined inertias of the motor and the load to accelerate in accordance with Equation 2.22:

$$\frac{d}{dt}\omega_m = \frac{T_J}{J_{eq}} \tag{2.23}$$

where the net torque $T_J = T_{em} - T_L$ and the equivalent combined inertia $J_{eq} = J_M + J_L$.

Example 2.3

In Figure 2.6a, each structure has the same inertia as the cylinder in Example 2.2. The load torque T_L is negligible. Calculate the required electromagnetic torque, if the speed is to increase linearly from rest to 1,800 rpm in 5 s.

Solution   Using the results of Example 2.2, the combined inertia of the system is

$$J_{eq} = 2 \times 0.029 = 0.058 \text{ kg} \cdot \text{m}^2$$

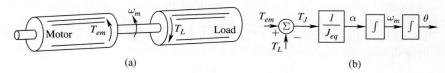

(a)   (b)

FIGURE 2.6   Motor and load torque interaction with a rigid coupling.

The angular acceleration is

$$\frac{d}{dt}\omega_m = \frac{\Delta\omega_m}{\Delta t} = \frac{(1800/60)2\pi}{5} = 37.7 \text{ rad/s}^2$$

Therefore, from Equation 2.23,

$$T_{em} = 0.058 \times 37.7 = 2.19 \text{ Nm}$$

Equation 2.23 shows that the net torque is the quantity that causes acceleration, which in turn leads to changes in speed and position. Integrating the acceleration $\alpha(t)$ with respect to time,

$$\text{Speed } \omega_m(t) = \omega_m(0) + \int_0^t \alpha(\tau) \, d\tau \qquad (2.24)$$

where $\omega_m(0)$ is the speed at $t=0$ and τ is a variable of integration. Further integrating $\omega_m(t)$ in Equation 2.24 with respect to time yields

$$\theta(t) = \theta(0) + \int_0^t \omega_m(\tau)d\tau \qquad (2.25)$$

where $\theta(0)$ is the position at $t = 0$, and τ is again a variable of integration. Equations 2.23 through 2.25 indicate that torque is the fundamental variable for controlling speed and position. Equations 2.23 through 2.25 can be represented in a block-diagram form, as shown in Figure 2.6b.

Example 2.4
Consider that the rotating system shown in Figure 2.6a, with the combined inertia $J_{eq} = 2 \times 0.029 = 0.058 \text{ kg·m}^2$, is required to have the angular speed profile shown in Figure 2.1b. The load torque is zero. Calculate and plot, as functions of time, the electromagnetic torque required from the motor and the change in position.

Solution   In the plot of Figure 2.1b, the magnitude of the acceleration and the deceleration is 100 rad/s². During the intervals of acceleration and deceleration, since $T_L = 0$,

$$T_{em} = T_J = J_{eq}\frac{d\omega_m}{dt} = \pm 5.8 \text{ Nm}$$

as shown in Figure 2.7.
During intervals with a constant speed, no torque is required. Since the position θ is the time-integral of speed, the resulting change of position (assuming that the initial position is zero) is also plotted in Figure 2.7.
In a rotational system shown in Figure 2.8, if a net torque T causes the cylinder to rotate by a differential angle $d\theta$, the differential work done is

$$dW = T \, d\theta \qquad (2.26)$$

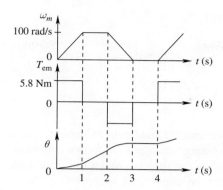

FIGURE 2.7   Speed, torque and angle variations with time.

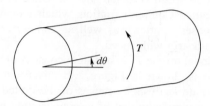

FIGURE 2.8   Torque, work and power.

If this differential rotation takes place in a differential time dt, the power can be expressed as

$$p = \frac{dW}{dt} = T\frac{d\theta}{dt} = T\,\omega_m \tag{2.27}$$

where $\omega_m = d\theta/dt$ is the angular speed of rotation. Substituting for T from Equation 2.21 into Equation 2.27,

$$p = J\frac{d\omega_m}{dt}\omega_m \tag{2.28}$$

Integrating both sides of Equation 2.28 with respect to time, assuming that the speed ω_m and the kinetic energy W at time $t = 0$ are both zero, the kinetic energy stored in the rotating mass of inertia J is

$$W = \int_0^t p(\tau)\,d\tau = J\int_0^t \omega_m\frac{d\omega_m}{d\tau}\,d\tau = J\int_0^{\omega_m} \omega_m\,d\omega_m = \tfrac{1}{2}J\omega_m^2 \tag{2.29}$$

This stored kinetic energy can be recovered by making the power $p(t)$ reverse direction, that is, by making $p(t)$ negative.

Example 2.5
In Example 2.3, calculate the kinetic energy stored in the combined inertia at a speed of 1,800 rpm.

Solution   From Equation 2.29,

$$W = \frac{1}{2}(J_L + J_M)\omega_m^2 = \frac{1}{2}(0.029 + 0.029)\left(2\pi\frac{1800}{60}\right)^2 = 1030.4 \; J$$

2.4 FRICTION

Friction within the motor and the load acts to oppose rotation. Friction occurs in the bearings that support rotating structures. Moreover, moving objects in air encounter windage or drag. In vehicles, this drag is a major force that must be overcome. Therefore, friction and windage can be considered as opposing forces or torque that must be overcome. The frictional torque is generally nonlinear in nature. We are all familiar with the need for a higher force (or torque) in the beginning (from rest) to set an object in motion. This friction at zero speed is called stiction. Once in motion, the friction may consist of a component called Coulomb friction, which remains independent of speed magnitude (it always opposes rotation), as well as another component called viscous friction, which increases linearly with speed.

In general, the frictional torque T_f in a system consists of all of the aforementioned components. An example is shown in Figure 2.9; this friction characteristic may be linearized for an approximate analysis by means of the dotted line. With this approximation, the characteristic is similar to that of viscous friction in which

$$T_f = B\omega_m \tag{2.30}$$

where B is the coefficient of viscous friction or viscous damping.

Example 2.6
The aerodynamic drag force in automobiles can be estimated as $f_L = 0.046 \, C_w A u^2$, where the coefficient 0.046 has the appropriate units, the drag force is in N, C_w is the drag coefficient (a unit-less quantity), A is the vehicle cross-sectional area in m^2, and u is the sum of the vehicle speed and headwind in km/h [4]. If A = 1.8 m^2 for two vehicles with $C_w = 0.3$ and $C_w = 0.5$ respectively, calculate the drag force and the power required to overcome it at the speeds of 50 km/h, and 100 km/h.

Solution   The drag force is $f_L = 0.046 \, C_w A u^2$ and the power required at the constant speed, from Equation 2.6, is $P = f_L u$ where the speed is expressed in m/s. Table 2.1 lists

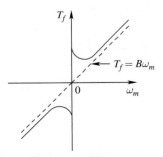

FIGURE 2.9   Actual and linearized friction characteristics.

TABLE 2.1 The drag force and the power required

Vehicle	$u = 50$ km/h		$u = 100$ km/h	
$C_w = 0.3$	$f_L = 62\text{-}06$ N	$P = 0.86$ kW	$f_L = 248.2$ N	$P = 6.9$ kW
$C_w = 0.5$	$f_L = 103.4$ N	$P = 1.44$ kW	$f_L = 413.7$ N	$P = 11.5$ kW

the drag force and the power required at various speeds for the two vehicles. Since the drag force F_L depends on the square of the speed, the power depends on the cube of the speed.

Traveling at 50 km/h compared to 100 km/h requires 1/8th the power, but it takes twice as long to reach the destination. Therefore, the energy required at 50 km/h would be 1/4th that at 100 km/h.

2.5 TORSIONAL RESONANCES

In Figure 2.6, the shaft connecting the motor with the load was assumed to be of infinite stiffness, that is, the two were rigidly connected. In reality, any shaft will twist (flex) as it transmits torque from one end to the other. In Figure 2.10, the torque T_{shaft} available to be transmitted by the shaft is

$$T_{shaft} = T_{em} - J_M \frac{d\omega_m}{dt} \qquad (2.31)$$

This torque at the load-end overcomes the load torque and accelerates it,

$$T_{shaft} = T_L + J_L \frac{d\omega_L}{dt} \qquad (2.32)$$

The twisting or the flexing of the shaft, in terms of the angles at the two ends, depends on the shaft torsional or the compliance coefficient K:

$$(\theta_M - \theta_L) = \frac{T_{shaft}}{K} \qquad (2.33)$$

where θ_M and θ_L are the angular rotations at the two ends of the shaft. If K is infinite, $\theta_M = \theta_L$. For a shaft of finite compliance, these two angles are not equal, and the shaft acts as a spring. This compliance in the presence of energy stored in the masses and inertias of the system can lead to resonance conditions at certain frequencies. This phenomenon is often termed torsional resonance. Such resonances should be avoided or kept low, otherwise they can lead to fatigue and failure of the mechanical components.

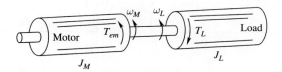

FIGURE 2.10 Motor and load-torque interaction with a rigid coupling.

2.6   ELECTRICAL ANALOGY

An analogy with electrical circuits can be very useful when analyzing mechanical systems. A commonly used analogy, though not a unique one, is to relate mechanical and electrical quantities as shown in Table 2.2.

For the mechanical system shown in Figure 2.10, Figure 2.11a shows the electrical analogy, where each inertia is represented by a capacitor from its node to a reference (ground) node. In this circuit, we can write equations similar to Equations 2.31 through 2.33. Assuming that the shaft is of infinite stiffness, the inductance representing it becomes zero, and the resulting circuit is shown in Figure 2.11b, where $\omega_m = \omega_M = \omega_L$. The two capacitors representing the two inertias can now be combined to result in a single equation similar to Equation 2.23.

Example 2.7
In an electric-motor drive similar to that shown in Figure 2.6a, the combined inertia is $J_{eq} = 5 \times 10^{-3}$ kg·m^2. The load torque opposing rotation is mainly due to friction, and can be described as $T_L = 0.5 \times 10^{-3} \omega_L$. Draw the electrical equivalent circuit and plot the electromagnetic torque required from the motor to bring the system linearly from rest to a speed of 100 rad/s in 4 s, and then to maintain that speed.

Solution   The electrical equivalent circuit is shown in Figure 2.12a. The inertia is represented by a capacitor of 5 mF, and the friction by a resistance $R = 1/(0.5 \times 10^{-3}) = 2000\ \Omega$. The linear acceleration is $100/4 = 25$ rad/s^2, which in the equivalent electrical

TABLE 2.2   Torque−Current Analogy

Mechanical System	Electrical System
Torque (T)	Current (i)
Angular speed (ω_m)	Voltage (v)
Angular displacement (θ)	Flux linkage (ψ)
Moment of inertia (J)	Capacitance (C)
Spring constant (K)	1/inductance (1/L)
Damping coefficient (B)	1/resistance (1/R)
Coupling ratio (n_M/n_L)	Transformer ratio (n_L/n_M)

Note: The coupling ratio is discussed later in this chapter.

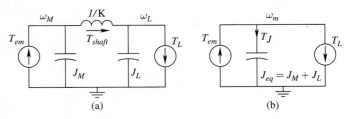

(a)   (b)

FIGURE 2.11   Electrical analogy: (a) Shaft of finite stiffness; (b) Shaft of infinite stiffness.

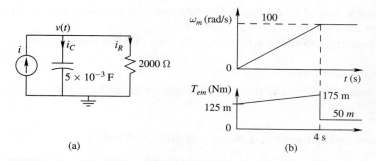

FIGURE 2.12 (a) Electrical equivalent; (b) Torque and speed variation.

circuit corresponds to $dv/dt = 25$ V/s. Therefore, during the acceleration period, $v(t) = 25t$. Thus, the capacitor current during the linear acceleration interval is

$$i_c(t) = C\frac{dv}{dt} = 125.0 \text{ mA} \quad 0 \leq t < 4 \text{ s} \tag{2.34a}$$

and the current through the resistor is

$$i_R(t) = \frac{v(t)}{R} = \frac{25\,t}{2000} = 12.5\,t \text{ mA} \quad 0 \leq t < 4 \text{ s} \tag{2.34b}$$

Therefore,

$$T_{em}(t) = (125.0 + 12.5\,t) \times 10^{-3} \text{ Nm} \quad 0 \leq t < 4 \text{ s} \tag{2.34c}$$

Beyond the acceleration stage, the electromagnetic torque is required only to overcome friction, which equals 50×10^{-3} Nm, as plotted in Figure 2.12b.

2.7 COUPLING MECHANISMS

Wherever possible, it is preferable to couple the load directly to the motor, to avoid the additional cost of the coupling mechanism and of the associated power losses. In practice, coupling mechanisms are often used for the following reasons:

- A rotary motor is driving a load that requires linear motion
- The motors are designed to operate at higher rotational speeds (to reduce their physical size) compared to the speeds required of the mechanical loads
- The axis of rotation needs to be changed

There are various types of coupling mechanisms. For conversion between rotary and linear motions, it is possible to use conveyor belts (belt and pulley), rack-and-pinion or a lead-screw type of arrangement. For rotary-to-rotary motion, various types of gear mechanisms are employed.

The coupling mechanisms have the following disadvantages:

- Additional power loss
- Introduction of nonlinearity due to a phenomenon called backlash
- Wear and tear

2.7.1 Conversion between Linear and Rotary Motion

In many systems, a linear motion is achieved by using a rotating-type motor, as shown in Figure 2.13.

In such a system, the angular and the linear speeds are related by the radius r of the drum:

$$u = r \, \omega_m \tag{2.35}$$

To accelerate the mass M in Figure 2.13 in the presence of an opposing force f_L, the force f applied to the mass, from Equation 2.1, must be

$$f = M \frac{du}{dt} + f_L \tag{2.36}$$

This force is delivered by the motor in the form of a torque T, which is related to f, using Equation 2.35, as

$$T = r \cdot f = r^2 M \frac{d\omega_m}{dt} + r f_L \tag{2.37}$$

Therefore, the electromagnetic torque required from the motor is

$$T_{em} = J_M \frac{d\omega_m}{dt} + \underbrace{r^2 M \frac{d \, \omega_m}{dt} + r f_L}_{\text{due to load}} \tag{2.38}$$

Example 2.8

In the vehicle of Example 2.6 with $C_w = 0.5$, assume that each wheel is powered by its own electric motor that is directly coupled to it. If the wheel diameter is 60 cm, calculate

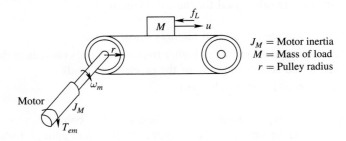

J_M = Motor inertia
M = Mass of load
r = Pulley radius

FIGURE 2.13   Combination of rotary and linear motion.

the torque and the power required from each motor to overcome the drag force, when the vehicle is traveling at a speed of 100 km/h.

Solution   In Example 2.6, the vehicle with $C_w = 0.5$ presented a drag force $f_L = 413.7$ N at the speed $u = 100$ km/h. The force required from each of the four motors is $f_M = \frac{f_L}{4} = 103.4$ N. Therefore, the torque required from each motor is

$$T_M = f_M \, r = 103.4 \times \frac{0.6}{2} = 31.04 \text{ Nm}$$

From Equation 2.35,

$$\omega_m = \frac{u}{r} = \left(\frac{100 \times 10^3}{3600}\right) \frac{1}{(0.6/2)} = 92.6 \text{ rad/s}$$

Therefore, the power required from each motor is

$$T_M \omega_m = 2.87 \text{ kW}$$

2.7.2   Gears

For matching speeds, Figure 2.14 shows a gear mechanism where the shafts are assumed to be of infinite stiffness and the masses of the gears are ignored. We will further assume that there is no power loss in the gears. Both gears must have the same linear speed at the point of contact. Therefore, their angular speeds are related by their respective radii r_1 and r_2 such that

$$r_1 \omega_M = r_2 \omega_L \tag{2.39}$$

and

$$\omega_M T_1 = \omega_L T_2 \quad \text{(assuming no power loss)} \tag{2.40}$$

Combining Equations 2.39 and 2.40,

$$\frac{r_1}{r_2} = \frac{\omega_L}{\omega_M} = \frac{T_1}{T_2} \tag{2.41}$$

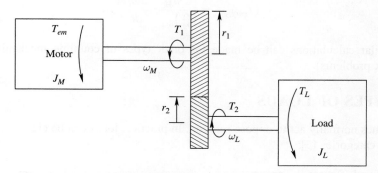

FIGURE 2.14   Gear mechanism for coupling the motor to the load.

where T_1 and T_2 are the torques at the ends of the gear mechanism, as shown in Figure 2.14. Expressing T_1 and T_2 in terms of T_{em} and T_L in Equation 2.41,

$$\underbrace{\left(T_{em} - J_M \frac{d\omega_M}{dt}\right)}_{T_1} \frac{\omega_M}{\omega_L} = \underbrace{\left(T_L + J_L \frac{d\omega_L}{dt}\right)}_{T_2} \tag{2.42}$$

From Equation 2.42, the electromagnetic torque required from the motor is

$$T_{em} = \underbrace{\left[J_M + \left(\frac{\omega_L}{\omega_M}\right)^2 J_L\right]}_{J_{eq}} \frac{d\omega_M}{dt} + \left(\frac{\omega_L}{\omega_M}\right) T_L \qquad \left(\text{note: } \frac{d\omega_L}{dt} = \frac{d\omega_M}{dt}\frac{\omega_L}{\omega_M}\right) \tag{2.43}$$

where the equivalent inertia at the motor side is

$$J_{eq} = J_M + \left(\frac{\omega_L}{\omega_M}\right)^2 J_L = J_M + \left(\frac{r_1}{r_2}\right)^2 J_L \tag{2.44}$$

2.7.2.1 Optimum Gear Ratio

Equation 2.43 shows that the electromagnetic torque required from the motor to accelerate a motor-load combination depends on the gear ratio. In a basically inertial load where T_L can be assumed to be negligible, T_{em} can be minimized, for a given load-acceleration $\frac{d\omega_L}{dt}$, by selecting an optimum gear ratio $(r_1/r_2)_{opt.}$. The derivation of the optimum gear ratio shows that the load inertia "seen" by the motor should equal the motor inertia, that is, in Equation 2.44

$$J_M = \left(\frac{r_1}{r_2}\right)_{opt.}^2 J_L \quad or \quad \left(\frac{r_1}{r_2}\right)_{opt.} = \sqrt{\frac{J_M}{J_L}} \tag{2.45a}$$

and, consequently,

$$J_{eq} = 2 J_M \tag{2.45b}$$

With the optimum gear ratio, in Equation 2.43, using $T_L = 0$, and using Equation 2.41,

$$(T_{em})_{opt.} = \frac{2 J_M}{\left(\frac{r_1}{r_2}\right)_{opt.}} \frac{d\omega_L}{dt} \tag{2.46}$$

Similar calculations can be made for other types of coupling mechanisms (see homework problems).

2.8   TYPES OF LOADS

Load torques normally act to oppose rotation. In practice, loads can be classified into the following categories [5]:

1. Centrifugal (Squared) Torque
2. Constant Torque

3. Squared Power
4. Constant Power

Centrifugal loads such as fans and blowers require torque that varies with speed^2 and load power that varies with speed^3. In constant-torque loads such as conveyors, hoists, cranes, and elevators, torque remains constant with speed, and the load power varies linearly with speed. In squared-power loads such as compressors and rollers, torque varies linearly with speed and the load power varies with speed^2. In constant-power loads such as winders and unwinders, the torque beyond a certain speed range varies inversely with speed, and the load power remains constant with speed.

2.9 FOUR-QUADRANT OPERATION

In many high-performance systems, drives are required to operate in all four quadrants of the torque-speed plane, as shown in Figure 2.15b.

The motor drives the load in the forward direction in quadrant 1, and in the reverse direction in quadrant 3. In both of these quadrants, the average power is positive and flows from the motor to the mechanical load. In order to control the load speed rapidly, it may be necessary to operate the system in the regenerative braking mode, where the direction of power is reversed so that it flows from the load into the motor, and usually into the utility (through the power-processing unit). In quadrant 2, the speed is positive, but the torque produced by the motor is negative. In quadrant 4, the speed is negative and the motor torque is positive.

2.10 STEADY STATE AND DYNAMIC OPERATIONS

As discussed in section 2.8, each load has its own torque-speed characteristic. For high-performance drives, in addition to the steady-state operation, the dynamic operation—how the operating point changes with time—is also important. The change of speed of the motor-load combination should be accomplished rapidly and without any oscillations (which otherwise may destroy the load). This requires a good design of the closed-loop controller, as discussed in Chapter 8, which deals with control of drives.

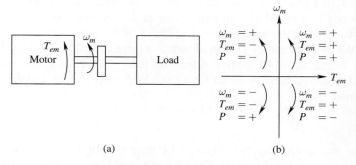

(a)   (b)

FIGURE 2.15 Four-quadrant requirements in drives.

SUMMARY/REVIEW QUESTIONS

1. What are the MKS units for force, torque, linear speed, angular speed, speed and power?
2. What is the relationship between force, torque, and power?
3. Show that torque is the fundamental variable in controlling speed and position.
4. What is the kinetic energy stored in a moving mass and a rotating mass?
5. What is the mechanism for torsional resonances?
6. What are the various types of coupling mechanisms?
7. What is the optimum gear ratio to minimize the torque required from the motor for a given load-speed profile as a function of time?
8. What are the torque-speed and the power-speed profiles for various types of loads?

REFERENCES

1. H. Gross (ed.), *Electric Feed Drives for Machine Tools* (New York: Siemens and Wiley, 1983).
2. *DC Motors and Control ServoSystem—An Engineering Handbook*, 5th ed. (Hopkins, MN: Electro-Craft Corporation, 1980).
3. M. Spong and M. Vidyasagar, *Robot Dynamics and Control* (New York: John Wiley & Sons, 1989).
4. Robert Bosch, *Automotive Handbook* (Robert Bosch GmbH, 1993).
5. T. Nondahl, Proceedings of the NSF/EPRI-Sponsored Faculty Workshop on "Teaching of Power Electronics," June 25–28, 1998, University of Minnesota.

PROBLEMS

2.1 A constant torque of 5 *Nm* is applied to an unloaded motor at rest at time $t = 0$. The motor reaches a speed of 1800 rpm in 3 *s*. Assuming the damping to be negligible, calculate the motor inertia.

2.2 Calculate the inertia if the cylinder in Example 2.2 is hollow, with the inner radius $r_2 = 4$ cm.

2.3 A vehicle of mass 1,500 kg is traveling at a speed of 50 km/hr. What is the kinetic energy stored in its mass? Calculate the energy that can be recovered by slowing the vehicle to a speed of 10 km/hr.

Belt-and-Pulley Systems

2.4 Consider the belt and pulley system in Fig 2.13. Inertias other than that shown in the figure are negligible. The pulley radius $r = 0.09$ m, and the motor inertia $J_M = 0.01$ kg·m^2. Calculate the torque T_{em} required to accelerate a load of 1.0 kg from rest to a speed of 1 m/s in a time of 4 *s*. Assume the motor torque to be constant during this interval.

2.5 For the belt and pulley system shown in Figure 2.13, M = 0.02 kg. For a motor with inertia $J_M = 40$ g·cm^2, determine the pulley radius that minimizes the torque required from the motor for a given load-speed profile. Ignore damping and the load force f_L.

Gears

2.6 In the gear system shown in Figure 2.14, the gear ratio $n_L/n_M = 3$ where n equals the number of teeth in a gear. The load and motor inertia are $J_L = 10$ kg·m^2 and

$J_M = 1.2$ kg·m^2. Damping and the load-torque T_L can be neglected. For the load-speed profile shown in Figure 2.1b, draw the profile of the electromagnetic torque T_{em} required from the motor as a function of time.

2.7 In the system of Problem 2.6, assume a triangular speed profile of the load with equal acceleration and deceleration rates (starting and ending at zero speed). Assuming a coupling efficiency of 100%, calculate the time needed to rotate the load by an angle of 30° if the magnitude of the electromagnetic torque (positive or negative) from the motor is 500 Nm.

2.8 The vehicle in Example 2.8 is powered by motors that have a maximum speed of 5000 rpm. Each motor is coupled to the wheel using a gear mechanism. (a) Calculate the required gear ratio if the vehicle's maximum speed is 150 km/hr, and (b) calculate the torque required from each motor at the maximum speed.

2.9 Consider the system shown in Figure 2.14. For $J_M = 40$ g·cm^2 and $J_L = 60$ g·cm^2, what is the optimum gear ratio to minimize the torque required from the motor for a given load-speed profile? Neglect damping and external load torque.

Lead-Screw Mechanism

2.10 Consider the lead-screw drive shown in Figure P2.10. Derive the following equation in terms of pitch s, where $\dot{u}_L=$ linear acceleration of the load, $J_M=$ motor inertia, $J_s =$ screw arrangement inertia, and the coupling ratio $n = \frac{s}{2\pi}$:

$$T_{em} = \frac{\dot{u}_L}{n}\left[J_M + J_s + n^2(M_T + M_W)\right] + nF_L$$

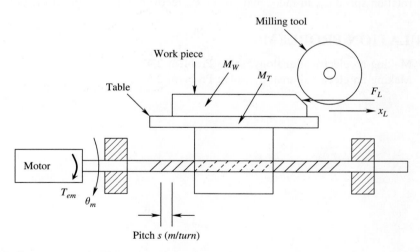

FIGURE P2.10 Lead-screw system.

Applications in Wind Turbines and Electric Vehicles

2.11 In wind-turbines, the shaft power available is given as follows, where the pitch-angle θ is nearly zero to "catch" all the wind energy available:

$$P_{shaft} = C_p\left(\frac{1}{2}\rho A_r V_W^3\right)$$

where C_p is the wind-turbine Coefficient of Performance (a unit-less quantity), ρ is the air density, A_r is the area swept by the rotor-blades, and V_W is the wind speed, all in MKS units. The rotational speed of the wind turbine is controlled such that it is operating near its optimum value of the coefficient of performance, with $C_p = 0.48$. Assume the combined efficiency of the gear-box, the generator and the power electronics converter to be 90%, and the air density to be $1.2 \ kg/m^3$. $A_r = 4,000 \ m^2$. Calculate the electrical power output of such a wind-turbine at its rated wind speed of 13 m/s.

2.12   A wind turbine is rotating at 22 rpm in steady state at a wind speed of 13 m/s, and producing 1.5 MW of power. The inertia of the mechanism is $3.4 \times 10^6 \ kg - m^2$. Suddenly there is a short-circuit on the electric grid, and the electrical output goes to zero for 2 seconds. Calculate the increase in speed in rpm during this interval. Assume that the shaft-torque remains constant and all other efficiencies to be 100% for the purpose of this calculation.

2.13   In an electric vehicle, each wheel is powered by its own motor. The vehicle weight is 2,000 kg. This vehicle increases in its speed linearly from 0 to 60 mph in 10 seconds. The tire diameter is 70 cm. Calculate the maximum power required from each motor in kW.

2.14   In an electric vehicle, each of the four wheels is supplied by its own motor. This EV weighs 1,000 kg, and the tire diameter is 50 cm. Using regenerative braking, its speed is brought from 20 m/s (72 kilometers per hour) to zero in 10 seconds, linearly with time. Neglect all losses. Calculate and plot, as a function of time for each wheel, the following: the electromagnetic deceleration torque T_{em} in Nm, rotation speed ω_m in rad/s, and power P_m recovered in kW. Label the plots.

SIMULATION PROBLEMS

2.15   Making an electrical analogy, solve Problem 2.4.
2.16   Making an electrical analogy, solve Problem 2.6.

3

REVIEW OF BASIC
ELECTRIC CIRCUITS

3.1 INTRODUCTION

The purpose of this chapter is to review elements of the basic electric circuit theory that are essential to the study of electric drives: the use of phasors to analyze circuits in sinusoidal steady state, the reactive power, the power factor, and the analysis of three-phase circuits.

In this book, we will use MKS units and the IEEE-standard letters and graphic symbols whenever possible. The lowercase letters v and i are used to represent instantaneous values of voltages and currents that vary as functions of time. They may or may not be shown explicitly as functions of time t. A current's positive direction is indicated by an arrow, as shown in Figure 3.1. Similarly, the voltage polarities must be indicated. The voltage v_{ab} refers to the voltage of node "a" with respect to node "b," thus $v_{ab} = v_a - v_b$.

3.2 PHASOR REPRESENTATION IN SINUSOIDAL
STEADY STATE

In linear circuits with sinusoidal voltages and currents of frequency f applied for a long time to reach steady state, all circuit voltages and currents are at a frequency $f(=\omega/2\pi)$. To analyze such circuits, calculations are simplified by means of phasor-domain analysis. The use of phasors also provides a deeper insight (with relative ease) into circuit behavior.

In the phasor domain, the time-domain variables $v(t)$ and $i(t)$ are transformed into phasors which are represented by the complex variables $\overline{V}$ and $\overline{I}$. Note that these phasors

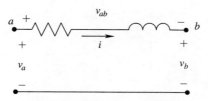

FIGURE 3.1 Conventions for currents and voltages.

31

are expressed by uppercase letters with a bar "-a" on top. In a complex (real and imaginary) plane, these phasors can be drawn with a magnitude and an angle.

A co-sinusoidal time function is taken as a reference phasor; for example, the voltage expression in Equation 3.1 below is represented by a phasor, which is entirely real with an angle of zero degrees:

$$v(t) = \hat{V} \cos \omega t \quad \Leftrightarrow \quad \overline{V} = \hat{V} \angle 0 \tag{3.1}$$

Similarly,

$$i(t) = \hat{I} \cos (\omega t - \phi) \quad \Leftrightarrow \quad \overline{I} = \hat{I} \angle -\phi \tag{3.2}$$

where "^" indicates the peak amplitude. These voltage and current phasors are drawn in Figure 3.2. We should note the following in Equations 3.1 and 3.2: we have chosen the peak values of voltages and currents to represent the phasor magnitudes, and the frequency ω is implicitly associated with each phasor. Knowing this frequency, a phasor expression can be re-transformed into a time-domain expression.

Using phasors, we can convert differential equations into easily solved algebraic equations containing complex variables. Consider the circuit of Figure 3.3a in a sinusoidal steady state with an applied voltage at a frequency $f \, (= \omega/2\pi)$. In order to calculate the current in this circuit, remaining in time domain, we would be required to solve the following differential equation:

$$R i(t) + L \frac{di(t)}{dt} + \frac{1}{C} \int i(t) \cdot dt = \hat{V} \cos(\omega t) \tag{3.3}$$

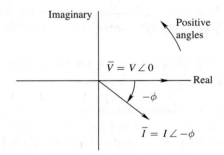

FIGURE 3.2 Phasor diagram.

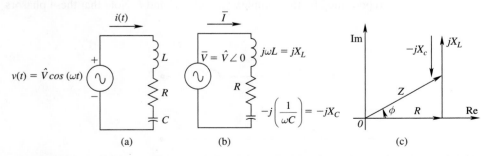

FIGURE 3.3 (a) Time domain circuit; (b) Phasor domain circuit; (c) Impedance triangle.

Using phasors, we can redraw the circuit of Figure 3.3a in Figure 3.3b, where the inductance L is represented by $j\omega L$ and the capacitance C is signified by $-j(\frac{1}{\omega C})$. In the phasor-domain circuit, the impedance Z of the series-connected elements is obtained by the impedance triangle of Figure 3.3c as

$$Z = R + jX_L - jX_c \tag{3.4}$$

where

$$X_L = \omega L, \quad \text{and} \quad X_c = \frac{1}{\omega C} \tag{3.5}$$

This impedance can be expressed as

$$Z = |Z| \angle \phi \tag{3.6a}$$

where

$$|Z| = \sqrt{R^2 + \left(\omega L - \frac{1}{\omega C}\right)^2} \quad \text{and} \quad \phi = \tan^{-1}\left[\frac{\left(\omega L - \frac{1}{\omega C}\right)}{R}\right] \tag{3.6b}$$

It is important to recognize that while Z is a complex quantity, it is *not* a phasor and does *not* have a corresponding time-domain expression.

Example 3.1

Calculate the impedance seen from the terminals of the circuit in Figure 3.4 under a sinusoidal steady state at a frequency $f = 60\text{Hz}$.

Solution

$$Z = j0.1 + (-j5.0 \| 2.0)$$

$$Z = j0.1 + \frac{-j10}{(2-j5)} = 1.72 - j0.59 = 1.82 \angle -18.9° \ \Omega$$

Using the impedance in Equation 3.6, the current in Figure 3.3b can be obtained as

$$\bar{I} = \frac{\bar{V}}{Z} = \left(\frac{\hat{V}}{|Z|}\right) \angle -\phi \tag{3.7}$$

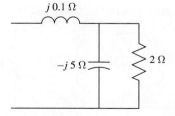

FIGURE 3.4 Impedance network.

where $\hat{I} = \frac{\hat{V}}{|Z|}$ and ϕ is as calculated from Equation 3.6b. Using Equation 3.2, the current can be expressed in the time domain as

$$i(t) = \frac{\hat{V}}{|Z|} \cos(\omega t - \phi) \tag{3.8}$$

In the impedance triangle of Figure 3.3c, a positive value of the phase angle ϕ implies that the current lags behind the voltage in the circuit of Figure 3.3a. Sometimes, it is convenient to express the inverse of the impedance, which is called admittance:

$$Y = \frac{1}{Z} \tag{3.9}$$

The phasor-domain procedure for solving $i(t)$ is much easier than solving the differential-integral equation given by Equation 3.3 (see homework problems 3.3 and 3.4).

Example 3.2
Calculate the current $\overline{I}_1$ and $i_1(t)$ in the circuit of Figure 3.5 if the applied voltage has an rms value of 120 V and a frequency of 60 Hz. Assume $\overline{V}_1$ to be the reference phasor.

Solution   For an rms value of 120 V, the peak amplitude is $\hat{V}_1 = \sqrt{2} \times 120 = 169.7$ V. With $\overline{V}_1$ as the reference phasor, it can be written as $\overline{V}_1 = 169.7 \angle 0°$ V. Impedance of the circuit seen from the applied voltage terminals is

$$Z = (R_1 + jX_1) + (jX_m)\|(R_2 + jX_2)$$

$$= (0.3 + j0.5) + \frac{(j15)(7 + j0.2)}{(j15) + (7 + j0.2)} = (5.92 + j3.29) = 6.78 \angle 29° \ \Omega$$

$$\overline{I}_1 = \frac{\overline{V}_1}{Z} = \frac{169.7 \angle 0°}{6.78 \angle 29°} = 25.0 \angle -29° \ A$$

Therefore,

$$i_1(t) = 25.0 \cos(\omega t - 29°) A.$$

The rms value of this current is $25.0/\sqrt{2} = 17.7 A$.

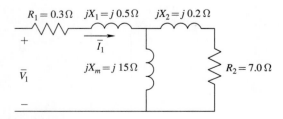

FIGURE 3.5   Example 3.2.

3.2.1 Power, Reactive Power, and Power Factor

Consider the generic circuit of Figure 3.6 in a sinusoidal steady state. Each subcircuit may consist of passive (R-L-C) elements and active voltage and current sources. Based on the arbitrarily chosen voltage polarity and the current direction shown in Figure 3.6, the instantaneous power $p(t) = v(t)i(t)$ is delivered by subcircuit 1 and absorbed by subcircuit 2.

This is because in subcircuit 1 the positively-defined current is coming out of the positive-polarity terminal (the same as in a generator). On the other hand, the positively-defined current is entering the positive-polarity terminal in subcircuit 2 (the same as in a load). A negative value of $p(t)$ reverses the roles of subcircuit 1 and subcircuit 2.

Under a sinusoidal steady-state condition at a frequency f, the complex power S, the reactive power Q, and the power factor express how "effectively" the real (average) power P is transferred from one subcircuit to the other.

If $v(t)$ and $i(t)$ are in phase, $p(t) = v(t)i(t)$, as shown in Figure 3.7a, pulsates at twice the steady-state frequency. But, at all times, $p(t) \geq 0$, and therefore the power always flows in one direction: from subcircuit 1 to subcircuit 2. Now consider the waveforms of Figure 3.7b, where the $i(t)$ waveform lags behind the $v(t)$ waveform by a phase angle $\phi(t)$. Now, $p(t)$ becomes negative during a time interval of (ϕ/ω) during each half-cycle. A negative instantaneous power implies power flow in the opposite direction. This back-and-forth flow of power indicates that the real (average) power is not optimally transferred from one subcircuit to the other, as is the case in Figure 3.7a.

The circuit of Figure 3.6 is redrawn in Figure 3.8a in the phasor domain. The voltage and the current phasors are defined by their magnitudes and phase angles as

$$\overline{V} = \hat{V} \angle \phi_v \qquad \text{and} \qquad \overline{I} = \hat{I} \angle \phi_i \qquad (3.10)$$

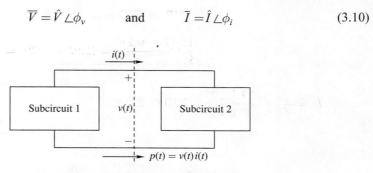

FIGURE 3.6 A generic circuit divided into two subcircuits.

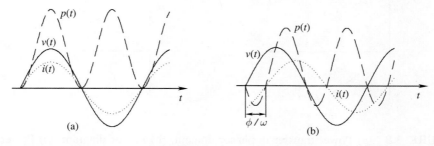

FIGURE 3.7 Instantaneous power with sinusoidal currents and voltages.

In Figure 3.8b, it is assumed that $\phi_v = 0$ and that ϕ_i has a negative value. To express real, reactive, and complex powers, it is convenient to use the rms voltage value V and the rms current value I, where

$$V = \frac{1}{\sqrt{2}} \hat{V} \qquad \text{and} \qquad I = \frac{1}{\sqrt{2}} \hat{I} \qquad (3.11)$$

The complex power S is defined as

$$S = \frac{1}{2} \overline{V} \overline{I}^* (* \text{ indicates complex conjugate}) \qquad (3.12)$$

Therefore, substituting the expressions for voltage and current into Equation 3.12, and noting that $\overline{I}^* = \hat{I} \angle -\phi_i$, in terms of the rms values of Equation 3.11,

$$S = V \angle \quad \phi_v I \angle -\phi_i = V I \angle (\phi_v - \phi_i) \qquad (3.13)$$

The difference between the two phase angles is defined as

$$\phi = \phi_v - \phi_i \qquad (3.14)$$

Therefore,

$$S = V I \angle \phi = P + jQ \qquad (3.15)$$

where

$$P = V I \cos \phi \qquad (3.16)$$

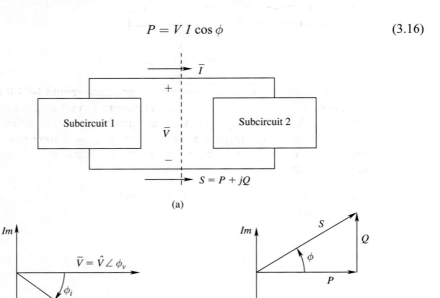

(a)

(b)   (c)

FIGURE 3.8   (a) Power transfer in phasor domain; (b) Phasor diagram; (c) Power triangle.

and

$$Q = V I \sin \phi \qquad (3.17)$$

The power triangle corresponding to Figure 3.8b is shown in Figure 3.8c. From Equation 3.15, the magnitude of S, also called the "apparent power," is

$$|S| = \sqrt{P^2 + Q^2} \qquad (3.18)$$

and

$$\phi = \tan^{-1}\left(\frac{Q}{P}\right) \qquad (3.19)$$

The above quantities have the following units: P: W (Watts); Q: *Var* (Volt-Amperes Reactive) assuming by convention that an inductive load draws positive vars; $|S|$: *VA* (Volt-amperes); finally, ϕ_v, ϕ_i, ϕ: radians, measured positively in a counter-clockwise direction with respect to the reference axis (drawn horizontally from left to right).

The physical significance of the apparent power $|S|$, P, and Q should be understood. The cost of most electrical equipment such as generators, transformers, and transmission lines is proportional to $|S|(= \text{VI})$, since their electrical insulation level and the magnetic core size depend on the voltage V, and the conductor size depends on the current I. The real power P has physical significance since it represents the useful work being performed plus the losses. In most situations, it is desirable to have the reactive power Q be zero.

To support the above discussion, another quantity called the power factor is defined. The power factor is a measure of how effectively a load draws real power:

$$\text{power factor} = \frac{P}{|S|} = \frac{P}{VI} = \cos \phi \qquad (3.20)$$

which is a dimension-less quantity. Ideally, the power factor should be 1.0 (that is, Q should be zero) in order to draw real power with a minimum current magnitude and hence minimize losses in electrical equipment and transmission and distribution lines. An inductive load draws power at a lagging power factor where the current lags behind the voltage. Conversely, a capacitive load draws power at a leading power factor where the load current leads the load voltage.

Example 3.3
Calculate P, Q, S, and the power factor of operation at the terminals in the circuit of Figure 3.5 in Example 3.2. Draw the power triangle.

Solution

$$P = V_1 I_1 \cos \phi = 120 \times 17.7 \cos 29° = 1857.7 \ W$$
$$Q = V_1 I_1 \sin \phi = 120 \times 17.7 \times \sin 29° = 1029.7 \ \text{VAR}$$
$$|S| = V_1 I_1 = 120 \times 17.7 = 2124 \ \text{VA}$$

From Equation 3.19, $\phi = \tan^{-1}\frac{Q}{P} = 29°$. The power triangle is shown in Figure 3.9. Note that the angle S in the power triangle is the same as the impedance angle ϕ in Example 3.2.

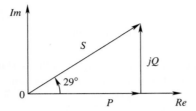

FIGURE 3.9   Power Triangle

Also note the following for the inductive impedance in the above example: (1) The impedance is $Z = |Z| \angle \phi$, where ϕ is positive. (2) The current lags the voltage by the impedance angle ϕ. This corresponds to a lagging power factor of operation. (3) In the power triangle, the impedance angle ϕ relates P, Q, and S. (4) An inductive impedance, when applied a voltage, draws a positive reactive power (vars). If the impedance were to be capacitive, the phase angle ϕ would be negative, and the impedance would draw a negative reactive power (in other words, the impedance would supply a positive reactive power).

3.3   THREE-PHASE CIRCUITS

Basic understanding of three-phase circuits is just as important in the study of electric drives as in power systems. Nearly all electricity is generated by means of three-phase ac generators. Figure 3.10 shows a one-line diagram of a three-phase transmission and distribution system. Generated voltages (usually between 22 and 69 kV) are stepped up by means of transformers to 230 kV to 500 kV level for transferring power over transmission lines from the generation site to load centers. Most motor loads above a few kW in power rating operate from three-phase voltages. In most ac motor drives, the input to the drive may be a single-phase or a three-phase line-frequency ac. However, motors are almost always supplied by three-phase, adjustable frequency ac, with the exception of the small, two-phase fan motors used in electronic equipment.

The most common configurations of three-phase ac circuits are wye-connections and delta connections. We will investigate both of these under sinusoidal steady-state conditions. In addition, we will assume a balanced condition, which implies that all three

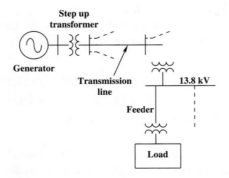

FIGURE 3.10   One-line diagram of a three-phase transmission and distribution system.

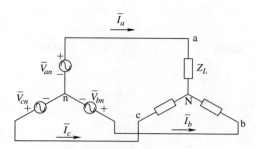

FIGURE 3.11 Y-connected source and load.

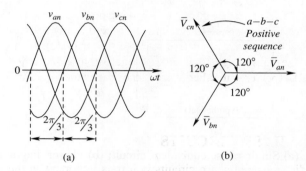

FIGURE 3.12 Three-phase voltages in time and phasor domain.

voltages are equal in magnitude and displaced by 120° ($2\pi/3$ radians) with respect to each other. Consider the wye-connected source and the load shown in the phasor domain in Figure 3.11.

The phase sequence is commonly assumed to be $a - b - c$, which is considered a positive sequence. In this sequence, the phase "a" voltage leads the phase "b" voltage by 120°, and phase "b" leads phase "c" by 120° ($2\pi/3$ radians), as shown in Figure 3.12. This applies to both the time domain and the phasor domain. Notice that in the $a - b - c$ sequence voltages plotted in Figure 3.12a, first v_{an} reaches its positive peak, and then v_{bn} reaches its positive peak $2\pi/3$ radians later, and so on. We can represent these voltages in the phasor form as

$$\overline{V}_{an} = \hat{V}_s \angle 0°, \qquad \overline{V}_{bn} = \hat{V}_s \angle -120°, \qquad \text{and} \qquad \overline{V}_{cn} = \hat{V}_s \angle -240° \qquad (3.21)$$

where $\hat{V}_s$ is the phase-voltage amplitude and the phase "a" voltage is assumed to be the reference (with an angle of zero degrees). For a balanced set of voltages given by Equation 3.21, at any instant, the sum of these phase voltages equals zero:

$$\overline{V}_{an} + \overline{V}_{bn} + \overline{V}_{cn} = 0 \quad \text{and} \quad v_{an}(t) + v_{bn}(t) + v_{cn}(t) = 0 \qquad (3.22)$$

3.3.1 Per-Phase Analysis

A three-phase circuit can be analyzed on a per-phase basis, provided that it has a balanced set of source voltages and equal impedances in each of the phases. Such a circuit was

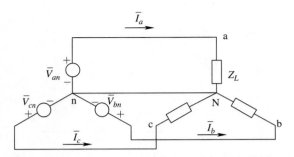

FIGURE 3.13   Hypothetical wire connecting source and load neutrals.

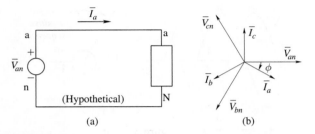

FIGURE 3.14   (a) Single-phase equivalent circuit; (b) Phasor diagram.

shown in Figure 3.11. In such a circuit, the source neutral "n" and the load neutral "N" are at the same potential. Therefore, "hypothetically" connecting these with a zero impedance wire, as shown in Figure 3.13, does not change the original three-phase circuit, which can now be analyzed on a per-phase basis.

Selecting phase "a" for this analysis, the per-phase circuit is shown in Figure 3.14a. If $Z_L = |Z_L| \angle \phi$, using the fact that in a balanced three-phase circuit, phase quantities are displaced by 120° with respect to each other, we find that

$$\bar{I}_a = \frac{\bar{V}_{an}}{Z_L} = \frac{\hat{V}_s}{|Z_L|} \angle - \phi,$$

$$\bar{I}_b = \frac{\bar{V}_{bn}}{Z_L} = \frac{\hat{V}_s}{|Z_L|} \angle (-\frac{2\pi}{3} - \phi), \text{ and} \tag{3.23}$$

$$\bar{I}_c = \frac{\bar{V}_{cn}}{Z_L} = \frac{\hat{V}_s}{|Z_L|} \angle (-\frac{4\pi}{3} - \phi)$$

The three-phase voltages and currents are shown in Figure 3.14b. The total real and reactive powers in a balanced three-phase circuit can be obtained by multiplying the per-phase values by a factor of 3. The power factor is the same as its per-phase value.

Example 3.4
In the balanced circuit of Figure 3.11, the rms phase voltages equal 120 V and the load impedance $Z_L = 5 \angle 30° \Omega$. Calculate the power factor of operation and the total real and reactive power consumed by the three-phase load.

Solution Since the circuit is balanced, only one of the phases, for example phase "*a*," needs to be analyzed:

$$\overline{V}_{an} = \sqrt{2} \times 120 \angle 0° \text{ V}$$

$$\overline{I}_a = \frac{\overline{V}_{an}}{Z_L} = \frac{\sqrt{2} \times 120 \angle 0°}{5 \angle 30°} = \sqrt{2} \times 24 \angle -30° \text{A}$$

The rms value of the current is 24 A. The power factor can be calculated as

$$\text{power factor} = \cos 30° = 0.866 \text{ (lagging)}$$

The total real power consumed by the load is

$$P = 3 V_{an} I_a \cos \phi = 3 \times 120 \times 24 \times \cos 30° = 7482 \text{ W}.$$

The total reactive power "consumed" by the load is

$$Q = 3 V_{an} I_a \sin \phi = 3 \times 120 \times 24 \times \sin 30° = 4320 \text{ VAR}.$$

3.3.2 Line-to-Line Voltages

In the balanced wye-connected circuit of Figure 3.11, it is often necessary to consider the line-to-line voltages, such as those between phases "*a*" and "*b*," and so on. Based on the previous analysis, we can refer to both neutral points "*n*" and "*N*" by a common term "*n*," since the potential difference between *n* and *N* is zero. Thus, in Figure 3.11,

$$\overline{V}_{ab} = \overline{V}_{an} - \overline{V}_{bn}, \quad \overline{V}_{bc} = \overline{V}_{bn} - \overline{V}_{cn}, \quad \text{and} \quad \overline{V}_{ca} = \overline{V}_{cn} - \overline{V}_{an} \tag{3.24}$$

as shown in the phasor diagram of Figure 3.15. Either using Equation 3.24, or graphically from Figure 3.15, we can show that

$$\overline{V}_{ab} = \sqrt{3} \hat{V}_s \angle \frac{\pi}{6}$$

$$\overline{V}_{bc} = \sqrt{3} \hat{V}_s \angle \left(\frac{\pi}{6} - \frac{2\pi}{3}\right) = \sqrt{3} \hat{V}_s \angle -\frac{\pi}{2} \tag{3.25}$$

$$\overline{V}_{ca} = \sqrt{3} \hat{V}_s \angle \left(\frac{\pi}{6} - \frac{4\pi}{3}\right) = \sqrt{3} \hat{V}_s \angle -\frac{7\pi}{6}$$

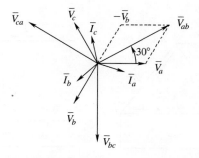

FIGURE 3.15 Line-to-line voltages in a balanced system.

Comparing Equations 3.21 and 3.25, we see that the line-to-line voltages have an amplitude of $\sqrt{3}$ times the phase voltage amplitude:

$$\hat{V}_{LL} = \sqrt{3}\,\hat{V}_s \tag{3.26}$$

and $\overline{V}_{ab}$ leads $\overline{V}_{an}$ by $\pi/6$ radians (30°).

3.3.3 Delta-Connected Loads

In ac-motor drives, the three motor phases may be connected in a delta configuration. Therefore, we will consider the circuit of Figure 3.16 where the load is connected in a delta configuration. Under a totally balanced condition, it is possible to replace the delta-connected load with an equivalent wye-connected load similar to that in Figure 3.11. We can then apply a per-phase analysis using Figure 3.14.

Consider the delta-connected load impedances of Figure 3.17a in a three-phase circuit.

In terms of the currents drawn, these are equivalent to the wye-connected impedances of Figure 3.17b, where

$$Z_y = \frac{Z_\Delta}{3} \tag{3.27}$$

The wye-connected equivalent circuit in Figure 3.17b is easy to analyze on a per phase basis.

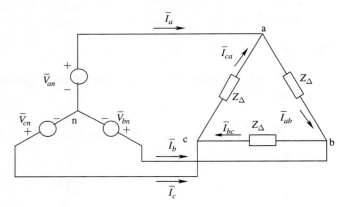

FIGURE 3.16 Delta-connected load.

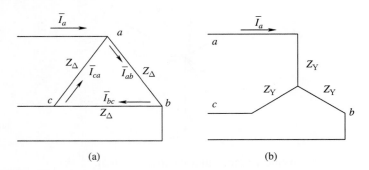

FIGURE 3.17 Delta-wye transformation.

SUMMARY/REVIEW QUESTIONS

1. Why is it important to always indicate the directions of currents and the polarities of voltages?
2. What are the meanings of $i(t), \hat{I}, I$, and $\bar{I}$?
3. In a sinusoidal waveform voltage, what is the relationship between the peak and the rms values?
4. How are currents, voltages, resistors, capacitors, and inductors represented in the phasor domain? Express and draw the following as phasors, assuming both ϕ_v and ϕ_i to be positive:

$$v(t) = \hat{V} \cos (\omega t + \phi_v) \qquad \text{and} \qquad i(t) = \hat{I} \cos (\omega t + \phi_i)$$

5. How is the current flowing through impedance $|Z| \angle \phi$ related to the voltage across it, in magnitude and phase?
6. What are real and reactive powers? What are the expressions for these in terms of rms values of voltage and current and the phase difference between the two?
7. What is complex power S? How are real and reactive powers related to it? What are the expressions for S, P, and Q, in terms of the current and voltage phasors? What is the power triangle? What is the polarity of the reactive power drawn by an inductive/capacitive circuit?
8. What are balanced three-phase systems? How can their analyses be simplified? What is the relation between line-to-line and phase voltages in terms of magnitude and phase? What are wye and delta connections?

REFERENCE

Any introductory textbook on Electric Circuits.

PROBLEMS

3.1   Calculate the rms values of currents with the waveforms shown in Figure P3.1.

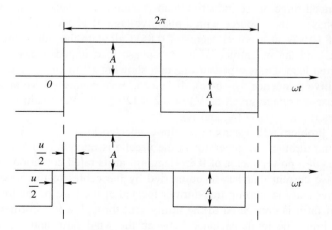

FIGURE P3.1   Current waveforms.

3.2   Calculate the rms values of currents with the waveforms shown in Figure P3.1.

3.3   Express the following voltages as phasors: (a) $v_1(t) = \sqrt{2} \times 120 \cos (\omega t\text{-}30°)$V and (b) $v_2(t) = \sqrt{2} \times 120 \cos (\omega t + 30°)$V.

3.4   The series R-L-C circuit of Figure 3.3a is in a sinusoidal steady state at a frequency of 60 Hz. $V = 120$ V, $R = 1.3\ \Omega$, $L = 20$ mH, and $C = 100\ \mu$F. Calculate $i(t)$ in this circuit by solving the differential equation Equation 3.3.

3.5   Repeat Problem 3.3 using the phasor-domain analysis.

3.6   In a linear circuit in sinusoidal steady-state with only one active source $\overline{V} = 90 \angle 30°$ V, the current in a branch is $\overline{I} = 5 \angle 15°$ A. Calculate the current in the same branch if the source voltage were to be $120 \angle 0°$ V.

3.7   In the circuit of Figure 3.5 in Example 3.2, show that the real and reactive powers supplied at the terminals equal the sum of their individual components, that is $P = \sum_k I_k^2 R_k$ and $Q = \sum_k I_k^2 X_k$.

3.8   An inductive load connected to a 120-V (rms), 60-Hz ac source draws 1 kW at a power factor of 0.8. Calculate the capacitance required in parallel with the load in order to bring the combined power factor to 0.95 (lagging).

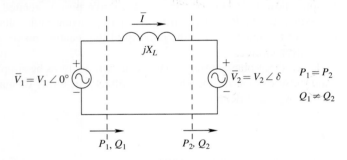

FIGURE P3.2   Power flow with AC sources.

3.9   In the circuit of Figure P3.2, $\overline{V}_1 = \sqrt{2} \times 120 \angle 0$ V and $X_L = 0.5\ \Omega$. Show $\overline{V}_1, \overline{V}_2,$ and $\overline{I}$ on a phasor diagram and calculate P_1 and Q_1 for the following values of $\overline{I}$: (a) $\sqrt{2} \times 10 \angle 0$ A, (b) $\sqrt{2} \times 10 \angle 180°$A, (c) $\sqrt{2} \times 10 \angle 90°$A, and (d) $\sqrt{2} \times 10 \angle -90°$A.

3.10   A balanced three-phase inductive load is supplied in steady state by a balanced three-phase voltage source with a phase voltage of 120 V rms. The load draws a total of 10 kW at a power factor of 0.85. Calculate the rms value of the phase currents and the magnitude of the per-phase load impedance, assuming a wye-connected load. Draw a phasor diagram showing all three voltages and currents.

3.11   A positive sequence (a–b–c), balanced, wye-connected voltage source has the phase-a voltage given as $\overline{V}_a = \sqrt{2} \times 120 \angle\ 30°$ V. Obtain the time-domain voltages $v_a(t)$, $v_b(t)$, $v_c(t)$, and $v_{ab}(t)$.

3.12   Repeat Problem 3.9, assuming a delta-connected load.

3.13   In a wind-turbine, the generator in the nacelle is rated at 690 V and 2.3 MW. It operates at a power factor of 0.85 (lagging) at its rated conditions. Calculate the per-phase current that has to be carried by the cables to the power electronics converter and the step-up transformer located at the base of the tower.

3.14   A wind farm is connected to the utility grid through a transformer and a distribution line. The total reactance between the wind farm and the utility grid is

$X_T = 0.2$ pu. The voltage at the wind farm is $\overline{V}_{WF} = 1.0 \angle 0°\ pu$. This wind farm is producing power $P = 1\ pu$ and supplying the reactive power of $Q = 0.1\ pu$ from the point of its interconnection to the rest of the system. Calculate the magnitude of the voltage at the grid and the current supplied from the wind farm. Draw the phasor diagram showing the voltages and current.

3.15   In Problem 3.14, assume that the magnitude of the grid voltage is 1.0 pu and the wind farm voltage is also to be maintained at 1 pu, when the wind farm is producing $P = 1\ pu$. Calculate the reactive power Q that the wind farm should supply from the point of its interconnection to the rest of the system.

3.16   In the per-phase diagram shown in Figure P3.3a, the grid voltage is $\overline{V}_s = 1.0 \angle 0°\ pu$. $\overline{V}_{conv}$ represents the ac voltage that can be synthesized with the appropriate magnitude and phase to obtain the desired current $\overline{I}$, as shown in the phasor diagram of Figure P3.3b. The reactance between the two voltage sources is $X = 0.05\ pu$. (a) Calculate $\overline{V}_{conv}$ for obtaining the following values of $\overline{I}$: $\overline{I} = 1.0 \angle -30°\ pu$, $\overline{I} = 1.0 \angle 30°\ pu$, $\overline{I} = -1.0 \angle -30°\ pu$, and $\overline{I} = -1.0 \angle 30°\ pu$. (b) for each value of $\overline{I}$ in part (a), calculate the real power P and the reactive power Q supplied to the grid voltage by $\overline{V}_{conv}$.

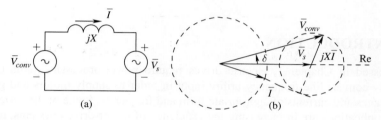

(a)   (b)

FIGURE P3.3

SIMULATION PROBLEMS

3.17   Repeat Problem 3.3 in sinusoidal steady state by means of computer simulation.
3.18   Repeat Problem 3.9 in sinusoidal steady state by means of computer simulation.
3.19   Repeat Problem 3.11 in sinusoidal steady state by means of computer simulation.

4

BASIC UNDERSTANDING OF SWITCH-MODE POWER ELECTRONIC CONVERTERS IN ELECTRIC DRIVES

4.1 INTRODUCTION

As discussed in Chapter 1, electric drives require power-processing units (PPUs) to efficiently convert line-frequency utility input in order to supply motors and generators with voltages and currents of appropriate form and frequency. Some of the sustainability-related applications are in increasing the efficiency of motor-driven systems, harnessing of energy in the wind, and electric transportation of various types, as discussed in Chapter 1. Similar to linear amplifiers, power-processing units amplify the input control signals. However, unlike linear amplifiers, PPUs in electric drives use switch-mode power electronics principles to achieve high energy efficiency and low cost, size, and weight. In this chapter, we will examine the basic switch-mode principles, topologies, and control for the processing of electrical power in an efficient and controlled manner.

4.2 OVERVIEW OF POWER PROCESSING UNITS (PPUS)

The discussion in this chapter is excerpted from [1], to which the reader is referred for a systematic discussion. In many applications such as wind-turbines, the voltage-link structure of Figure 4.1 is used. To provide the needed functionality to the converters in Figure 4.1, the transistors and diodes, which can block voltage only of one polarity, has led to this commonly-used voltage-link-structure.

This structure consists of two separate converters, one on the utility side and the other on the load side. The dc ports of these two converters are connected to each other with a parallel capacitor forming a dc-link, across which the voltage polarity does not reverse, thus allowing unipolar voltage-blocking transistors to be used within these converters. The capacitor in parallel with the two converters forms a dc voltage-link; hence this structure is called a *voltage-link* (or a *voltage-source*) structure. This structure

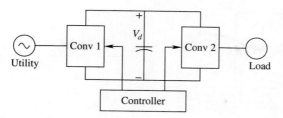

FIGURE 4.1 Voltage-link system.

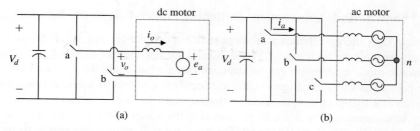

FIGURE 4.2 Switch mode converters for (a) dc- and (b) ac-machine drives.

is used in a very large power range, from a few tens of watts to several megawatts, even extending to hundreds of megawatts in utility applications. Therefore, we will mainly focus on this voltage-link structure in this book, although there are other structures as well. It should be remembered that the power flow in Figure 4.1 reverses when the role of the utility and the load is interchanged.

To understand how converters in Figure 4.1 operate, our emphasis will be to discuss how the load-side converter, with the dc voltage as input, synthesizes dc or low-frequency sinusoidal voltage outputs. Functionally, this converter operates as a linear amplifier, amplifying a control signal, dc in case of dc-motor drives, and ac in case of ac-motor drives. The power flow through this converter should be reversible.

The dc voltage V_d (assumed to be constant) is used as the input voltage to the load-side switch-mode converter in Figure 4.1. The task of this converter, depending on the machine type, is to deliver an adjustable-magnitude dc or sinusoidal ac to the machine by amplifying the signal from the controller by a constant gain. The power flow through the switch-mode converter must be able to reverse. In switch-mode converters, as their name implies, transistors are operated as switches: either fully on or fully off. The switch-mode converters used for dc- and ac-machine drives can be simply illustrated, as in Figures 4.2a and 4.2b, respectively, where each bi-positional switch constitutes a pole. The dc-dc converter for dc-machine drives in Figure 4.2a consists of two such poles, whereas the dc-to-three-phase ac converter shown in Figure 4.2b for ac-machine drives consists of three such poles.

Typically, PPU efficiencies exceed 95% and can exceed 98% in very large power ratings. Therefore, the energy efficiency of adjustable-speed drives is comparable to that of conventional line-fed motors; thus systems with adjustable-speed drives can achieve much higher overall *system* efficiencies (compared to their conventional counterparts) in many applications discussed in Chapter 1.

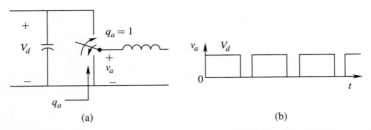

FIGURE 4.3   Switching power-pole as the building block in converters.

4.2.1   Switch-Mode Conversion: Switching Power-Pole as the Building Block

Achieving high energy-efficiency requires switch-mode conversion, where in contrast to linear power electronics, transistors (and diodes) are operated as switches, either on off. This switch-mode conversion can be explained by its basic building block, a switching power-pole a, as shown in Figure 4.3a. It effectively consists of a bi-positional switch, which forms a two-port: a voltage-port across a capacitor with a voltage V_d that cannot change instantaneously, and a current-port due the series inductor through which the current cannot change instantaneously. For now, we will assume the switch to be ideal with two positions: up or down, dictated by a switching signal q_a which takes on two values: 1 and 0, respectively.

The bi-positional switch "chops" the input dc voltage V_d into a train of high-frequency voltage pulses, shown by v_a waveform in Figure 4.3b, by switching up or down at a high repetition rate, called the switching frequency f_s. Controlling the pulse width within a switching cycle allows control over the switching-cycle-averaged value of the pulsed output, and this pulse-width modulation forms the basis of synthesizing adjustable dc and low-frequency sinusoidal ac outputs, as described in the next section. A switch-mode converter consists of one or more such switching power-poles.

4.2.2   Pulse-Width Modulation (PWM) of the Switching Power-Pole (constant f_s)

The objective of the switching power-pole, redrawn in Figure 4.4a, is to synthesize the output voltage such that its *switching-cycle-average* is of the desired value: dc or ac that varies sinusoidally at a low-frequency, compared to f_s. Switching at a constant switching-frequency f_s produces a train of voltage pulses in Figure 4.4b that repeat with a constant switching time-period T_s, equal to $1/f_s$.

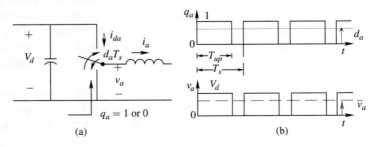

FIGURE 4.4   PWM of the switching power-pole.

Within each switching cycle with the time-period $T_s\ (= 1/f_s)$ in Figure 4.4b, the switching-cycle-averaged value $\bar{v}_a$ of the waveform is controlled by the pulse width T_{up} (during which the switch is in the up position and v_a equals V_d), as a ratio of T_s:

$$\bar{v}_a = \frac{T_{up}}{T_s} V_d = d_a V_d \qquad 0 \le d_a \le 1 \tag{4.1}$$

where $d_a (= T_{up}/T_s)$, which is the average of the q_a waveform shown in Figure 4.4b, is defined as the duty-ratio of the switching power-pole a, and the switching-cycle-averaged voltage is indicated by a "-" on top. The switching-cycle-averaged voltage and the switch duty-ratio are expressed by lowercase letters since they may vary as functions of time. The control over the switching-cycle-averaged value of the output voltage is achieved by adjusting or modulating the pulse width, which later on will be referred to as pulse-width-modulation (PWM). This switching power-pole and the control of its output by PWM set the stage for switch-mode conversion with high energy-efficiency.

We should note that $\bar{v}_a$ and d_a in the above discussion are discrete quantities and their values, calculated over a kth switching cycle, for example, can be expressed as $\bar{v}_{a,k}$ and $d_{a,k}$. However, in practical applications, the pulse-width T_{up} changes very slowly over many switching cycles, and hence we can consider them analog quantities as $\bar{v}_a(t)$ and $d_a(t)$ that are continuous functions of time. For simplicity, we may not show their time dependence explicitly.

4.2.3 Bi-Directional Switching Power-Pole

A bi-directional switching power-pole, through which the power flow can be in either direction, is implemented as shown in Figure 4.5a. In such a bi-directional switching power-pole, the positive inductor current i_L as shown in Figure 4.5b represents a Buck-mode of operation (where the power flow is from the higher voltage is to the lower voltage), where only the transistor and the diode associated with the Buck converter take part; the transistor conducts during $q = 1$, otherwise the diode conducts.

However, as shown in Figure 4.5c, the negative inductor current represents a Boost-mode of operation (where the power flow is from the lower voltage to the higher voltage), where only the transistor and the diode associated with the Boost converter take part; the transistor conducts during $q = 0$ $(q^- = 1)$, otherwise the diode conducts during $q = 1$ $(q^- = 0)$.

Figures 4.5b and c show that the combination of devices in Figure 4.5a renders it to be a switching power-pole that can carry i_L of either direction. This is shown as an

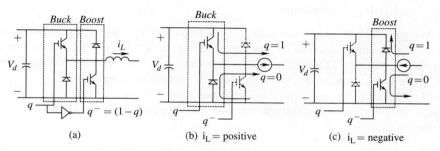

FIGURE 4.5 Bi-directional power flow through a switching power-pole.

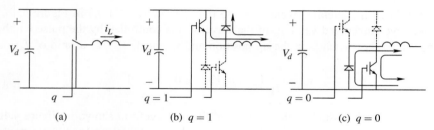

FIGURE 4.6   Bidirectional switching power-pole.

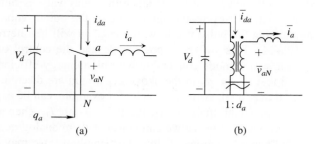

FIGURE 4.7   Switching-cycle averaged representation of the bi-directional power-pole.

equivalent switch in Figure 4.6a that is effectively in the "up" position when $q = 1$ as shown in Figure 4.6b, and in the "down" position when $q = 0$ as shown in Figure 4.6c, regardless of the direction of i_L.

The bi-directional switching power-pole of Figure 4.6a is repeated in Figure 4.7a for pole-a, with its switching signal identified as q_a. In response to the switching signal, it behaves as follows: "up" when $q_a = 1$; otherwise "down." Therefore, its switching-cycle averaged representation is an ideal transformer, shown in Figure 4.7b, with a turns-ratio $1 : d_a(t)$.

The switching-cycle averaged values of the variables at the voltage-port and the current-port in Figure 4.7b are related by $d_a(t)$ as follows:

$$\bar{v}_{aN} = d_a V_d \tag{4.2}$$

$$\bar{i}_{da} = d_a \bar{i}_a \tag{4.3}$$

4.2.4   Pulse-Width-Modulation (PWM) of the Bi-Directional Switching Power-Pole

The voltage of a switching power-pole at the current-port is always of a positive polarity. However, the output voltages of converters in Figure 4.2 for electric drives must be reversible in polarity. This is achieved by introducing a common-mode voltage in each power-pole as discussed below, and taking the differential output between the power-poles.

To obtain a desired switching-cycle averaged voltage $\bar{v}_{aN}$ in Figure 4.7b, that includes the common-mode voltage, requires the following power-pole duty-ratio from Equation 4.2:

$$d_a = \frac{\bar{v}_{aN}}{V_d} \tag{4.4}$$

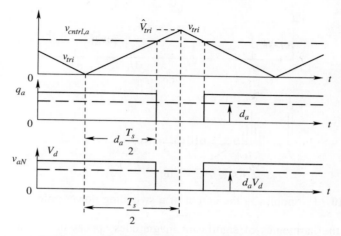

FIGURE 4.8 Waveforms for PWM in a switching power-pole.

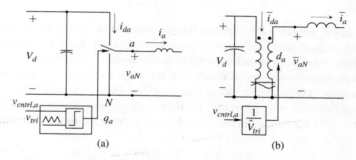

FIGURE 4.9 Switching power-pole and its duty-ratio control.

where V_d is the dc-bus voltage. To obtain the switching signal q_a to deliver this duty-ratio, a control voltage $v_{cntrl,a}$ is compared with a triangular-shaped carrier waveform of the switching-frequency f_s and amplitude $\hat{V}_{tri}$, as shown in Figure 4.8. Because of symmetry, only $T_s/2$, one-half the switching time-period, needs be considered. The switching-signal $q_a = 1$ if $v_{cntrl,a} > v_{tri}$; otherwise 0. Therefore in Figure 4.8,

$$v_{cntrl,a} = d_a \hat{V}_{tri} \tag{4.5}$$

The switching-cycle averaged representation of the switching power-pole in Figure 4.9a is shown by a controllable turns-ratio ideal transformer in Figure 4.9b, where the switching-cycle averaged representation of the duty-ratio control is in accordance with Equation 4.5.

The Fourier spectrum of the switching waveform v_{aN} is shown in Figure 4.10, which depends on the nature of the control signal. If the control voltage is dc, the output voltage has harmonics at the multiples of the switching frequency; that is at f_s, $2f_s$, etc as shown in Figure 4.10a. If the control voltage varies at a low-frequency f_1, as in electric

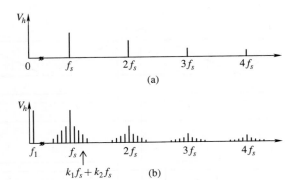

FIGURE 4.10   Harmonics in the output of a switching power-pole.

drives, then the harmonics of significant magnitudes appear in the side bands of the switching frequency and its multiples, as shown in Figure 4.10b, where

$$f_h = k_1 f_s + \underbrace{k_2 f_1}_{side\ bands} \tag{4.6}$$

in which k_1 and k_2 are constants that can take on values 1, 2, 3, and so on. Some of these harmonics associated with each power-pole are cancelled from the converter output voltages, where two or three such power-poles are used.

In the power-pole shown in Figure 4.9, the output voltage v_{aN} and its switching-cycle averaged $\bar{v}_{aN}$ are limited between 0 and V_d. To obtain an output voltage $\bar{v}_{an}$ (where "n" may be a fictitious node) that can become both positive and negative, a common-mode offset $\bar{v}_{com}$ is introduced in each power-pole so that the pole output voltage is

$$\bar{v}_{aN} = \bar{v}_{com} + \bar{v}_{an} \tag{4.7}$$

where $\bar{v}_{com}$ allows $\bar{v}_{an}$ to become both positive and negative around the common-mode voltage $\bar{v}_{com}$. In the differential output when two or three power-poles are used, the common-mode voltage gets eliminated.

4.3   CONVERTERS FOR DC MOTOR DRIVES $(-V_d < \bar{v}_o < V_d)$

Converters for dc motor drives, covered in Chapter 7, consist of two power-poles as shown in Figure 4.11a, where

$$\bar{v}_o = \bar{v}_{aN} - \bar{v}_{bN} \tag{4.8}$$

and $\bar{v}_o$ can assume both positive and negative values. Since the output voltage is desired to be in a full range, from $-V_d$ to $+V_d$, pole-a is assigned to produce $\bar{v}_o/2$ and pole-b is assigned to produce $-\bar{v}_o/2$ towards the output:

$$\bar{v}_{an} = \frac{\bar{v}_o}{2} \quad \text{and} \quad \bar{v}_{bn} = -\frac{\bar{v}_o}{2} \tag{4.9}$$

where "n" is a fictitious node as shown in Figure 4.11a, chosen to define the contribution of each pole towards $\bar{v}_o$.

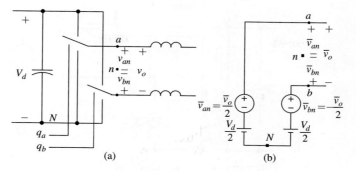

FIGURE 4.11 Converter for dc-motor drive.

To achieve equal excursions in positive and negative values of the switching-cycle averaged output voltage, the switching-cycle averaged common-mode voltage in each pole is chosen to be one-half the dc-bus voltage

$$\bar{v}_{com} = \frac{V_d}{2} \tag{4.10}$$

Therefore, from Equation 4.7

$$\bar{v}_{aN} = \frac{V_d}{2} + \frac{\bar{v}_o}{2} \quad \text{and} \quad \bar{v}_{bN} = \frac{V_d}{2} - \frac{\bar{v}_o}{2} \tag{4.11}$$

The switching-cycle averaged output voltages of the power-poles and the converter are shown in Figure 4.11b. From Equations 4.4 and 4.11

$$d_a = \frac{1}{2} + \frac{1}{2}\frac{\bar{v}_o}{V_d} \quad \text{and} \quad d_b = \frac{1}{2} - \frac{1}{2}\frac{\bar{v}_o}{V_d} \tag{4.12}$$

and from Equation 4.12

$$\bar{v}_o = (d_a - d_b)V_d \tag{4.13}$$

Example 4.1
In a dc-motor drive, the dc-bus voltage is $V_d = 350$ V. Determine the following: $\bar{v}_{com}$, $\bar{v}_{aN}$, and d_a for pole-a and similarly for pole-b, if the output voltage required is (a) $\bar{v}_0 = 300$ V and (b) $\bar{v}_0 = -300$ V.

Solution From Equation 4.10, $\bar{v}_{com} = \frac{V_d}{2} = 175$ V.

(a) For $\bar{v}_0 = 300$ V, from Equation 4.9, $\bar{v}_{an} = \bar{v}_o/2 = 150$ V and $\bar{v}_{bn} = -\bar{v}_o/2 = -150$ V. From Equation 4.11, $\bar{v}_{aN} = 325$ V and $\bar{v}_{bN} = 25$ V. From Equation 4.12, $d_a \simeq 0.93$ and $d_b \simeq 0.07$.

(b) For $\bar{v}_0 = -300$ V, $\bar{v}_{an} = \bar{v}_o/2 = -150$ V and $\bar{v}_{bn} = -\bar{v}_o/2 = 150$ V. Therefore from Equation 4.11, $\bar{v}_{aN} = 25$ V and $\bar{v}_{bN} = 325$ V. From Equation 4.12, $d_a \simeq 0.07$ and $d_b \simeq 0.93$.

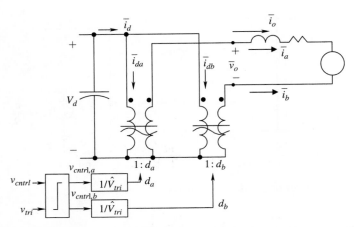

FIGURE 4.12 Switching-cycle averaged representation of the converter for dc drives.

The switching-cycle averaged representation of the two power-poles, along with the pulse-width-modulator, in a block-diagram form is shown in Figure 4.12.

In each power-pole of Figure 4.12, the switching-cycle averaged dc-side current is related to its output current by the pole duty-ratio

$$\bar{i}_{da} = d_a \bar{i}_a \qquad \text{and} \qquad \bar{i}_{db} = d_b \bar{i}_b \tag{4.14}$$

By Kirchhoff's current law, the total switching-cycle averaged dc-side current is

$$\bar{i}_d = \bar{i}_{da} + \bar{i}_{db} = d_a \bar{i}_a + d_b \bar{i}_b \tag{4.15}$$

Recognizing the directions with which the currents i_a and i_b are defined,

$$\bar{i}_a(t) = -\bar{i}_b(t) = \bar{i}_o(t) \tag{4.16}$$

Thus, substituting currents from Equation 4.15 into Equation 4.16

$$\bar{i}_d = (d_a - d_b)\bar{i}_o \tag{4.17}$$

Example 4.2

In the dc-motor drive of Example 4.1, the output current into the motor is $\bar{i}_o = 15$ A. Calculate the power delivered from the dc-bus and show that it is equals to the power delivered to the motor (assuming the converter to be lossless), if $\bar{v}_0 = 300$ V.

Solution Using the values for d_a and d_b from part (a) of Example 4.1, and $\bar{i}_o = 15$ A, from Equation 4.17, $\bar{i}_d(t) = 12.9$ A and therefore the power delivered by the dc-bus is $P_d = 4.515$ kW. Power delivered by the converter to the motor is $P_o = \bar{v}_o \bar{i}_o = 4.5$ kW, which is equal to the input power (neglecting the round-off errors).

Using Equations 4.5 and 4.12, the control voltages for the two poles are as follows:

$$v_{cntrl,a} = \frac{\hat{V}_{tri}}{2} + \frac{\hat{V}_{tri}}{2}\left(\frac{\bar{v}_o}{V_d}\right) \qquad \text{and} \qquad v_{cntrl,b} = \frac{\hat{V}_{tri}}{2} - \frac{\hat{V}_{tri}}{2}\left(\frac{\bar{v}_o}{V_d}\right) \tag{4.18}$$

FIGURE 4.13   Gain of the converter for dc drives.

In Equation 4.18, defining the second term in the two control voltages above as one-half the control voltage, that is,

$$\frac{v_{cntrl}}{2} = \frac{\hat{V}_{tri}}{2}\left(\frac{\bar{v}_o}{V_d}\right)$$

(4.19)

Equation 4.19 simplifies as follows

$$\bar{v}_o = \underbrace{\left(\frac{V_d}{\hat{V}_{tri}}\right)}_{k_{PWM}} v_{cntrl}$$

(4.20)

where $\left(V_d/\hat{V}_{tri}\right)$ is the converter gain k_{PWM}, from the feedback control signal to the switching-cycle averaged voltage output, as shown in Figure 4.13 in a block-diagram form.

4.3.1   Switching Waveforms in a Converter for DC Motor Drives

We will look further into the switching details of the converter in Figure 4.11a. To produce a positive output voltage, the control voltages are as shown in Figure 4.14. Only one-half the time-period, $T_s/2$, needs to be considered due to symmetry.

The pole output voltages v_{aN} and v_{bN} have the same waveform as the switching signals except for their amplitude. The output voltage v_o waveform shows that the effective switching frequency at the output is twice the original. That is, within the time-period of the switching-frequency f_s with which the converter devices are

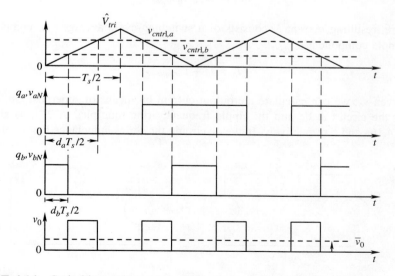

FIGURE 4.14   Switching voltage waveforms in a converter for dc drive.

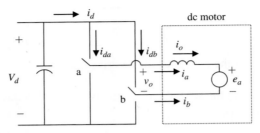

FIGURE 4.15   Currents defined in the converter for dc-motor drives.

switching, there are two complete cycles of repetition. Therefore, the harmonics in the output are at $(2f_s)$ and at its multiples. If the switching frequency is selected sufficiently large, the motor inductance may be enough to keep the ripple in the output current within acceptable range without the need for an external inductor in series.

Next, we will look at the currents associated with this converter, repeated in Figure 4.15. The pole currents $i_a = i_o$ and $i_b = -i_o$. The dc-side current $i_d = i_{da} + i_{db}$. The waveforms for these currents are shown by means of Example 4.3.

Example 4.3

In the dc-motor drive of Figure 4.15, assume the operating conditions are as follows: $V_d = 350$ V, $e_a = 236$ V(dc) and $\bar{i}_o = 4$ A. The switching-frequency f_s is 20 kHz. Assume that the series resistance R_a associated with the motor is 0.5 Ω. Calculate the series inductance L_a necessary to keep the peak-peak ripple in the output current to be 1.0 A at this operating condition. Assume that $\hat{V}_{tri} = 1$ V. Plot v_o, $\bar{v}_o$, i_o and i_d.

Solution   As seen from Figure 4.14, the output voltage v_o is a pulsating waveform that consists of a dc switching-cycle averaged $\bar{v}_o$ plus a ripple component $v_{o,ripple}$, which contains sub-components that are at very high frequencies (the multiples of $2f_s$):

$$v_o = \bar{v}_o + v_{o,ripple} \tag{4.21}$$

Therefore, resulting current i_o consists of a switching-cycle averaged dc component $\bar{i}_o$ and a ripple component $i_{o,ripple}$:

$$i_o = \bar{i}_o + i_{o,ripple} \tag{4.22}$$

For a given v_o, we can calculate the output current by means of superposition, by considering the circuit at dc and the ripple frequency (the multiples of $2f_s$), as shown in Figures 4.16a and b respectively. In the dc circuit, the series inductance has no effect and

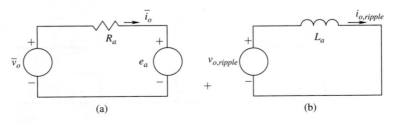

FIGURE 4.16   Superposition of dc and ripple-frequency variables.

hence is omitted from Figure 4.16a. In the ripple-frequency circuit of Figure 4.16b, the back-emf e_a, that is dc, is suppressed along with the series resistance R_a, which generally is negligible compared to the inductive reactance of L_a at very frequencies associated with the ripple.

From the circuit of Figure 4.16a,

$$\bar{v}_o = e_a + R_a \bar{i}_o = 238 \text{ V} \tag{4.23}$$

The switching waveforms are shown in Figure 4.17, which is based on Figure 4.14, where the details are shown for the first half-cycle. The output voltage v_o pulsates between 0 and $V_d = 350$ V, where from Equation 4.12, $d_a = 0.84$, and $d_b = 0.16$. At $f_s = 20$ kHz, $T_s = 50$ μs. Using Equations 4.21 and 4.23, the ripple voltage waveform is also shown in Figure 4.17, where during $\frac{d_a - d_b}{2} T_s$ (= 17.0μs), the ripple voltage in the circuit of

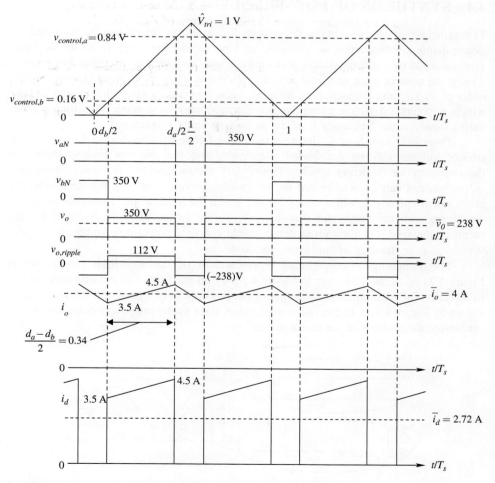

FIGURE 4.17 Switching current waveforms in Example 4.3.

Figure 4.16b is 112 V. Therefore, during this time-interval, the peak-to-peak ripple ΔI_{p-p} in the inductor current can be related to the ripple voltage as follows:

$$L_a \frac{\Delta I_{p-p}}{(d_a - d_b)T_s/2} = 112 \text{ V} \tag{4.24}$$

Substituting the values in the equation above with $\Delta I_{p-p} = 1$ A, $L_a = 1.9$ mH. As shown in Figure 4.17, the output current increases linearly during $(d_a - d_b)T_s/2$, and its waveform is symmetric around the switching-cycle averaged value; that is, it crosses the switching-cycle averaged value at the midpoint of this interval. The ripple waveform in other intervals can be found by symmetry. The dc-side current i_d flows only during $(d_a - d_b)T_s/2$ interval; otherwise it is zero as shown in Figure 4.17. Averaging over $T_s/2$, the switching-cycle averaged dc-side current $\bar{i}_d = 2.72$ A.

4.4 SYNTHESIS OF LOW-FREQUENCY AC

The principle of synthesizing a dc voltage for dc-motor drives can be extended for synthesizing low-frequency ac voltages, so long as the frequency f_1 of the ac being synthesized is two or three orders of magnitude smaller than the switching frequency f_s. This is the case in most ac-motor drives applications where f_1 is at 60 Hz (or is of the order of 60 Hz) and the switching frequency is a few tens of kHz. The control voltage, which is compared with a triangular waveform voltage to generate switching signals, varies slowly at the frequency f_1 of the ac voltage being synthesized.

Therefore, with $f_1 \ll f_s$, during a switching-frequency time-period $T_s(= 1/f_s)$, the control voltage can be considered pseudo-dc, and the analysis and synthesis for the converter for dc-drives applies. Figure 4.18 shows how the switching power-pole output voltage can be synthesized so, on switching-cycle averaged, it varies as shown at the low frequency f_1, where at any instant "under the microscope" shows the switching waveform with the duty ratio that depends on the switching-cycle averaged voltage being synthesized. The limit on switching-cycle averaged power-pole voltage is between 0 and V_d, as in the case of converters for dc drives.

The switching-cycle averaged representation of the switching power-pole in Figure 4.7a is, as shown earlier in Figure 4.7b, by an ideal transformer with controllable turns-ratio. The harmonics in the output of the power pole in a general form were shown earlier by Figure 4.10b. In the following section, three switching power-poles are used to synthesize three-phase ac for motor drives.

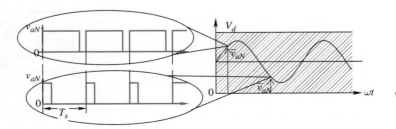

FIGURE 4.18 Waveforms of a switching power-pole to synthesize low-frequency ac.

4.5 THREE-PHASE INVERTERS

Converters for three-phase outputs consist of three power-poles as shown in Figure 4.19a. The switching-cycle averaged representation is shown in Figure 4.19b.

In Figure 4.19, $\bar{v}_{an}$, $\bar{v}_{bn}$ and $\bar{v}_{cn}$ are the desired balanced three-phase switching-cycle-averaged voltages to be synthesized: $\bar{v}_{an} = \hat{V}_{ph} \sin(\omega_1 t)$, $\bar{v}_{bn} = \hat{V}_{ph} \sin(\omega_1 t - 120°)$ and $\bar{v}_{cn} = \hat{V}_{ph} \sin(\omega_1 t - 240°)$. In series with these, common-mode voltages are added such that

$$\bar{v}_{aN} = \bar{v}_{com} + \bar{v}_{an} \quad \bar{v}_{bN} = \bar{v}_{com} + \bar{v}_{bn} \quad \bar{v}_{cN} = \bar{v}_{com} + \bar{v}_{cn} \tag{4.25}$$

These voltages are shown in Figure 4.20a. The common-mode voltages do not appear across the load; only $\bar{v}_{an}$, $\bar{v}_{bn}$, and $\bar{v}_{cn}$ appear across the load with respect to the load-neutral. This can be illustrated by applying the principle of superposition to the circuit of Figure 4.20a.

By "suppressing" $\bar{v}_{an}$, $\bar{v}_{bn}$ and $\bar{v}_{cn}$, only equal common-mode voltages are present in each phase, as shown in Figure 4.20b. If the current in one phase is i, then it will be the same in the other two phases. By Kirchhoff's current law at the load-neutral, $3i = 0$ and hence $i = 0$, and, therefore, the common-mode voltages do not appear across the load phases.

To obtain the switching-cycle averaged currents drawn from the voltage-port of each switching power-pole, we will assume the currents drawn by the motor load in

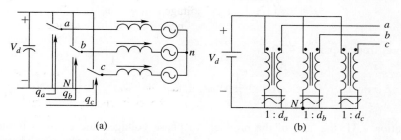

(a)   (b)

FIGURE 4.19   Three-phase converter.

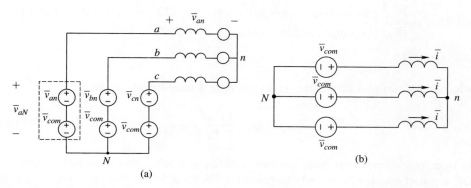

(a)

FIGURE 4.20   Switching-cycle averaged output voltages in a three-phase converter.

Figure 4.19b to be sinusoidal but lagging with respect to the switching-cycle averaged voltages in each phase by an angle ϕ_1, where $\bar{v}_{an}(t) = \hat{V}_{ph} \sin \omega_1 t$ and so on:

$$\bar{i}_a(t) = \hat{I} \sin(\omega_1 t - \phi_1), \ \bar{i}_b(t) = \hat{I} \sin(\omega_1 t - \phi_1 - 120°), \ \bar{i}_c(t) = \hat{I} \sin(\omega_1 t - \phi_1 - 240°)$$

(4.26)

Assuming that the ripple in the output currents is negligibly small, the average power output of the converter can be written as

$$P_o = \bar{v}_{aN}\bar{i}_a + \bar{v}_{bN}\bar{i}_b + \bar{v}_{cN}\bar{i}_c$$

(4.27)

Equating the average output power to the power input from the dc-bus and assuming the converter to be lossless

$$\bar{i}_d(t)V_d = \bar{v}_{aN}\bar{i}_a + \bar{v}_{bN}\bar{i}_b + \bar{v}_{cN}\bar{i}_c$$

(4.28)

Making use of Equation 4.26 into Equation 4.28,

$$\bar{i}_d(t)V_d = \bar{v}_{com}(\bar{i}_a + \bar{i}_b + \bar{i}_c) + \bar{v}_{an}\bar{i}_a + \bar{v}_{bn}\bar{i}_b + \bar{v}_{cn}\bar{i}_c$$

(4.29)

By Kirchhoff's current law at the load-neutral, the sum of all three phase currents within brackets in Equation 4.29 is zero

$$\bar{i}_a + \bar{i}_b + \bar{i}_c = 0$$

(4.30)

Therefore, from Equation 4.29,

$$\bar{i}_d(t) = \frac{1}{V_d}(\bar{v}_{an}\bar{i}_a + \bar{v}_{bn}\bar{i}_b + \bar{v}_{cn}\bar{i}_c)$$

(4.31)

In Equation 4.31, the sum of the products of phase voltages and currents is the three-phase power being supplied to the motor. Substituting for phase voltages and currents in Equation 4.31

$$\bar{i}_d(t) = \frac{\hat{V}_{ph}\hat{I}}{V_d}\left\{ \begin{array}{l} \sin(\omega_1 t) \sin(\omega_1 t - \phi_1) + \sin(\omega_1 t - 120°) \sin(\omega_1 t - \phi_1 - 120°) \\ + \sin(\omega_1 t - 240°) \sin(\omega_1 t - \phi_1 - 240°) \end{array} \right\}$$

(4.32)

which simplifies to a dc current, as it should, in a three-phase circuit:

$$\bar{i}_d(t) = I_d = \frac{3}{2}\frac{\hat{V}_{ph}\hat{I}}{V_d} \cos \phi_1$$

(4.33)

In three-phase converters there are two methods of synthesizing sinusoidal output voltages, out of which we will only discuss Sine-PWM. In Sine-PWM, the switching-cycle averaged output of power poles, $\bar{v}_{aN}$, $\bar{v}_{bN}$ and $\bar{v}_{cN}$, have a constant dc common-mode

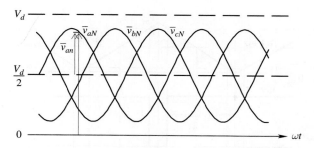

FIGURE 4.21 Switching-cycle averaged voltages due to Sine-PWM.

voltage $\bar{v}_{com} = \frac{V_d}{2}$, similar to that in dc-motor drives, around which $\bar{v}_{an}$, $\bar{v}_{bn}$, and $\bar{v}_{cn}$ can vary sinusoidally as shown in Figure 4.21:

$$\bar{v}_{aN} = \frac{V_d}{2} + \bar{v}_{an} \qquad \bar{v}_{bN} = \frac{V_d}{2} + \bar{v}_{bn} \qquad \bar{v}_{cN} = \frac{V_d}{2} + \bar{v}_{cn} \tag{4.34}$$

In Figure 4.21, using Equation 4.4, the plots of $\bar{v}_{aN}$, $\bar{v}_{bN}$, and $\bar{v}_{cN}$, each divided by V_d, are also the plots of d_a, d_b and d_c within the limits of 0 and 1:

$$d_a = \frac{1}{2} + \frac{\bar{v}_{an}}{V_d} \qquad d_b = \frac{1}{2} + \frac{\bar{v}_{bn}}{V_d} \qquad d_c = \frac{1}{2} + \frac{\bar{v}_{cn}}{V_d} \tag{4.35}$$

These power-pole duty-ratios define the turns-ratio in the ideal transformer representation of Figure 4.19b. As can be seen from Figure 4.21, at the limit, $\bar{v}_{an}$ can become a maximum of $\frac{V_d}{2}$ and hence the maximum allowable value of the phase-voltage peak is

$$(\hat{V}_{ph})_{max} = \frac{V_d}{2} \tag{4.36}$$

Therefore, using the properties of three-phase circuits where the line-line voltage magnitude is $\sqrt{3}$ times the phase-voltage magnitude, the maximum amplitude of the line-line voltage in Sine-PWM is limited to

$$(\hat{V}_{LL})_{max} = \sqrt{3}(\hat{V}_{ph})_{max} = \frac{\sqrt{3}}{2}V_d \simeq 0.867\, V_d \tag{4.37}$$

4.5.1 Switching Waveforms in a Three-Phase Inverter with Sine-PWM

In Sine-PWM, three sinusoidal control voltages equal the duty-ratios, given in Equation 4.35, multiplied by $\hat{V}_{tri}$. These are compared with a triangular waveform signal to generate the switching signals. These switching waveforms for Sine-PWM are shown by an example below.

Example 4.4
In a three-phase converter of Figure 4.19a, a Sine-PWM is used. The parameters and the operating conditions are as follows: $V_d = 350$ V, $f_1 = 60$ Hz, $\bar{v}_{an} = 160 \cos \omega_1 t$ volts etc,

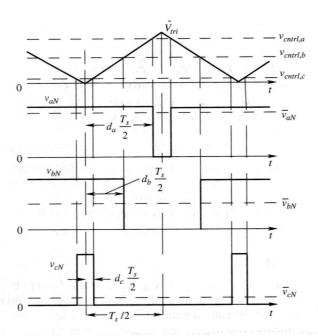

FIGURE 4.22   Switching waveforms in example 4.4.

and the switching frequency $f_s = 25$ kHz. $\hat{V}_{tri} = 1$ V. At $\omega_1 t = 15°$, calculate and plot the switching waveforms for one cycle of the switching frequency.

Solution   At $\omega_1 t = 15°$, $\bar{v}_{an} = 154\text{-}55$ V, $\bar{v}_{bn} = -41.41$ V and $\bar{v}_{cn} = -113.14$ V. Therefore, from Equation 4.34, $\bar{v}_{aN} = 329.55$ V, $\bar{v}_{bN} = 133.59$ V and $\bar{v}_{cN} = 61.86$ V. From Equation 4.35, the corresponding power-pole duty-ratios are $d_a = 0.942$, $d_b = 0.382$, and $d_c = 0.177$. For $\hat{V}_{tri} = 1$ V, these duty-ratios also equal the control voltages in volts. The switching time-period $T_s = 50$ μs. Based on this, the switching waveforms are shown in Figure 4.22.

It should be noted that there is another approach called Space-Vector PWM (SV-PWM), described in Reference [1], which fully utilizes the available dc-bus voltage, and results in the ac output that can be approximately 15% higher than that possible by using the Sine-PWM approach, both in a linear range where no lower-order harmonics appear. Sine-PWM is limited to $(\hat{V}_{LL})_{max} \simeq 0.867\, V_d$, as given by Equation 4.37, because it synthesizes output voltages on a per-pole basis, which does not take advantage of the three-phase properties. However, by considering line-line voltages, it is possible to get $(\hat{V}_{LL})_{max} = V_d$ in SV-PWM.

4.6   POWER SEMICONDUCTOR DEVICES [4]

Electric drives owe their market success, in part, to rapid improvements in power semiconductor devices and control ICs. Switch-mode power electronic converters require diodes and transistors, which are controllable switches that can be turned on and off by applying a small voltage to their gates. These power devices are characterized by the following quantities:

1. *Voltage Rating* is the maximum voltage that can be applied across a device in its *off*-state, beyond which the device "breaks down" and irreversible damage occurs.
2. *Current Rating* is the maximum current (expressed as instantaneous, average, and/or rms) that a device can carry, beyond which excessive heating within the device destroys it.
3. *Switching Speeds* are the speeds with which a device can make a transition from its *on*-state to *off*-state, or vice versa. Small switching times associated with fast-switching devices result in low switching losses, or considering it differently, fast-switching devices can be operated at high switching frequencies.
4. On-*State Voltage* is the voltage drop across a device during its on-state while conducting a current. The smaller this voltage is, the smaller the on-state power loss.

4.6.1 Device Ratings

Available power devices range in voltage ratings of several kV (up to 9 kV) and current ratings of several kA (up to 5 kA). Moreover, these devices can be connected in series and parallel to satisfy any voltage and current requirements. Their switching speeds range from a fraction of a microsecond to a few microseconds, depending on their other ratings. In general, higher-power devices switch more slowly than their low-power counterparts. The *on*-state voltage is usually in the range of 1 to 3 volts.

4.6.2 Power Diodes

Power diodes are available in voltage ratings of several kV (up to 9 kV) and current ratings of several kA (up to 5 kA). The on-state voltage drop across these diodes is usually of the order of 1 V. Switch-mode converters used in motor drives require fast-switching diodes. On the other hand, the diode rectification of line-frequency ac can be accomplished by slower-switching diodes, which have a slightly lower *on*-state voltage drop.

4.6.3 Controllable Switches

Transistors are controllable switches, which are available in several forms: Bipolar-Junction Transistors (BJTs), metal-oxide-semiconductor field-effect transistors (MOS-FETs), Gate Turn Off (GTO) thyristors, and insulated-gate bipolar transistor (IGBTs). In switch-mode converters for motor-drive applications, there are two devices that are primarily used: MOSFETs at low power levels and IGBTs in power ranges extending to MW levels. The following subsections provide a brief overview of their characteristics and capabilities.

4.6.3.1 MOSFETs

In applications at voltages below 200 volts and switching frequencies in excess of 50 kHz, MOSFETs are clearly the device of choice because of their low *on*-state losses in low-voltage ratings, their fast switching speeds, and their ease of control. The circuit symbol of an n-channel MOSFET is shown in Figure 4.23a. It consists of three terminals: drain (D), source (S), and gate (G). The main current flows between the drain and the source terminals. MOSFET *i-v* characteristics for various gate voltage values are shown

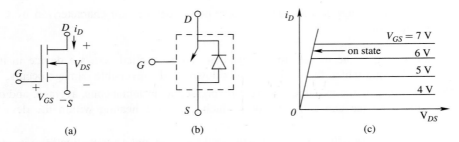

FIGURE 4.23   MOSFET characteristics.

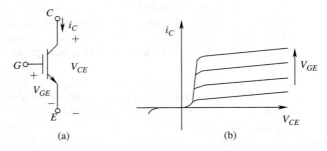

FIGURE 4.24   IGBT symbol and characteristics.

in Figure 4.23b; it is fully *off* and approximates an open switch when the gate-source voltage is zero. To turn the MOSFET *on* completely, a positive gate-to-source voltage, typically in a range of 10 to 15 volts, must be applied. This gate-source voltage should be continuously applied in order to keep the MOSFET in its *on*-state.

4.6.3.2   Insulated-Gate Bipolar Transistors (IGBTs)

IGBTs combine the ease of control of MOSFETs with low *on*-state losses, even at fairly high voltage ratings. Their switching speeds are sufficiently fast for switching frequencies up to 30 kHz. Therefore, they are used in a vast voltage and power range—from a fractional kW to many MW.

The circuit symbol for an IGBT is shown in Figure 4.24a and the *i-v* characteristics are shown in Figure 4.24b. Similar to MOSFETs, IGBTs have a high impedance gate, which requires only a small amount of energy to switch the device. IGBTs have a small *on*-state voltage, even in devices with large blocking-voltage ratings (for example, V_{on} is approximately 2 V in 1200-V devices). IGBTs can be designed to block negative voltages, but most commercially available IGBTs, by design to improve other properties, cannot block any appreciable reverse-polarity voltage (similar to MOSFETs).

Insulated-gate bipolar transistors have turn-on and turn-off times on the order of 1 microsecond and are available as modules in ratings as large as 3.3 kV and 1200 A. Voltage ratings of up to 5 kV are projected.

4.6.4   "Smart Power" Modules including Gate Drivers

A gate-drive circuitry, shown as a block in Figure 4.25, is required as an intermediary to interface the control signal coming from a microprocessor or an analog control integrated

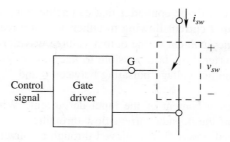

FIGURE 4.25   Block diagram of a gate-drive circuit.

circuit (IC) to the power semiconductor switch. Such gate-drive circuits require many components, passive as well as active. An electrical isolation may also be needed between the control-signal circuit and the circuit in which the power switch is connected. The gate-driver ICs, which include all of these components in one package, have been available for some time.

Lately, "Smart Power" modules, also called Power Integrated Modules (PIMs), have become available. Smart power modules combine more than one power switch and diode, along with the required gate-drive circuitry, into a single module. These modules also include fault protection and diagnostics. Such modules immensely simplify the design of power electronic converters.

4.6.5   Cost of MOSFETs and IGBTs

As these devices evolve, their relative cost continues to decline, for example, approximately 0.25 \$/A for 600-V devices and 0.50 \$/A for 1200-V devices. Power modules for the 3 kV class of devices cost approximately 1 \$/A.

SUMMARY/REVIEW QUESTIONS

1. What is the function of PPUs?
2. What are the sub-blocks of PPUs?
3. Qualitatively, how does a switch-mode amplifier differ from a linear amplifier?
4. Why does operating transistors as switches result in much smaller losses compared to operating them in their linear region?
5. How is a bi-positional switch realized in a converter pole?
6. What is the gain of each converter pole?
7. How does a switch-mode converter pole approach the output of a linear amplifier?
8. What is the meaning of $\bar{v}_{aN}(t)$?
9. How is the pole output voltage made linearly proportional to the input control signal?
10. What is the physical significance of the duty-ratio, for example $d_a(t)$?
11. How is pulse-width-modulation (PWM) achieved, and what is its function?
12. Instantaneous quantities on the two sides of the converter pole, for example pole-a, are related by the switching signal $q_a(t)$. What relates the average quantities on the two sides?
13. What is the equivalent model of a switch-mode pole in terms of its average quantities?

14. How is a switch-mode dc-dc converter that can achieve an output voltage of either polarity and an output current flowing in either direction realized?
15. What is the frequency content of the output voltage waveform in dc-dc converters?
16. In a dc-drive converter, how is it possible to keep the ripple in the output current small, despite the output voltage pulsating between 0 and V_d, or 0 and $-V_d$, during each switching cycle?
17. What is the frequency content of the input dc current? Where does the pulsating ripple component of the dc-side current flow through?
18. How is bi-directional power flow achieved through a converter pole?
19. What makes the average of the dc-side current in a converter pole related to the average of the output current by its duty-ratio?
20. How are three-phase, sinusoidal ac output voltages synthesized from a dc voltage input?
21. What are the voltage and current ratings and the switching speeds of various power semiconductor devices?

REFERENCES

1. N. Mohan, *Power Electronics: A First Course* (New York: John Wiley & Sons, 2011).
2. N. Mohan, *Power Electronics: Computer Simulation, Analysis and Education using PSpice* (January 1998), www.mnpere.com.
3. N. Mohan, T. Undeland, and W. P. Robbins, *Power Electronics: Converters, Applications and Design*, 3rd ed. (New York: John Wiley & Sons, 2003).

PROBLEMS

4.1 In a switch-mode converter pole-*a* in Figure 4.4a, $V_d = 150$ V, $\hat{V}_{tri} = 5$ V, and $f_s = 20$ kHz. Calculate the values of the control signal $v_{cntrl, a}$ and the pole duty-ratio d_a during which the switch is in its top position, for the following values of the average output voltage: $\bar{v}_{aN} = 125$ V and $\bar{v}_{aN} = 50$ V.

4.2 In Problem 4.1, assume the $i_a(t)$ waveform to be dc with a magnitude of 10 A. Draw the waveform of $i_{da}(t)$ for the two values of $\bar{v}_{aN}$.

DC-DC CONVERTERS (FOUR-QUADRANT CAPABILITY)

4.3 A switch-mode dc-dc converter uses a PWM-controller IC, which has a triangular waveform signal at 25 kHz with $\hat{V}_{tri} = 3$ V. If the input dc source voltage $V_d = 150$ V, calculate the gain k_{PWM} of this switch-mode amplifier.

4.4 In a switch-mode dc-dc converter of Figure 4.11a, $\frac{v_{cntrl}}{\hat{V}_{tri}=0.8}$ with a switching frequency $f_s = 20$ kHz and $V_d = 150$ V. Calculate and plot the ripple in the output voltage $v_o(t)$.

4.5 A switch-mode dc-dc converter is operating at a switching frequency of 20 kHz, and $V_d = 150$ V. The average current being drawn by the dc motor is 8.0 A. In the equivalent circuit of the dc motor, $E_a = 100$ V, $R_a = 0.25$ Ω, and $L_a = 4$ mH, all in series. (a) Plot the output current and calculate the peak-to-peak ripple and (b) plot the dc-side current.

4.6 In Problem 4.5, the motor goes into regenerative braking mode. The average current being supplied by the motor to the converter during braking is 8.0 A.

Plot the voltage and current waveforms on both sides of this converter. Calculate the average power flow into the converter.

4.7 In Problem 4.5, calculate $\bar{i}_{da}$, $\bar{i}_{db}$, and $\bar{i}_d (= I_d)$.

4.8 Repeat Problem 4.5 if the motor is rotating in the reverse direction, with the same current draw and the same induced emf E_a value of the opposite polarity.

4.9 Repeat Problem 4.8 if the motor is braking while it has been rotating in the reverse direction. It supplies the same current and produces the same induced emf E_a value of the opposite polarity.

4.10 Repeat problem 4.5 if a bi-polar voltage switching is used in the dc-dc converter. In such a switching scheme, the two bi-positional switches are operated in such a manner that when switch-a is in the top position, switch-b is in its bottom position, and vice versa. The switching signal for pole-a is derived by comparing the control voltage (as in Problem 4.5) with the triangular waveform.

DC-TO-THREE-PHASE AC INVERTERS

4.11 Plot $d_a(t)$ if the output voltage of the converter pole-a is $\bar{v}_{aN}(t) = \frac{V_d}{2} + 0.85 \frac{V_d}{2} \sin(\omega_1 t)$, where $\omega_1 = 2\pi \times 60$ rad/s.

4.12 In the three-phase dc-ac inverter of Figure 4.19, $V_d = 300$ V, $\hat{V}_{tri} = 1$ V, $\bar{v}_{an}(t) = 90 \sin(\omega_1 t)$, and $f_1 = 45$ Hz. Calculate and plot $d_a(t), d_b(t), d_c(t)$, $\bar{v}_{aN}(t), \bar{v}_{bN}(t), \bar{v}_{cN}(t)$, and $\bar{v}_{an}(t), \bar{v}_{bn}(t)$, and $\bar{v}_{cn}(t)$.

4.13 In the balanced three-phase dc-ac inverter shown in Figure 4.19, the phase-a average output voltage is $\bar{v}_{an}(t) = \frac{V_d}{2} 0.75 \sin(\omega_1 t)$, where $V_d = 300$ V and $\omega_1 = 2\pi \times 45$ rad/s. The inductance L in each phase is 5 mH. The ac-motor internal voltage in phase-a can be represented as $e_a(t) = 106.14 \sin(\omega_1 t - 6.6°)$ V, assuming this internal voltage to be purely sinusoidal. (a) Calculate and plot $d_a(t)$, $d_b(t)$, and $d_c(t)$, (b) sketch $\bar{i}_a(t)$, and (c) sketch $\bar{i}_{da}(t)$.

4.14 In Problem 4.13, calculate and plot $\bar{i}_d(t)$, which is the average dc current drawn from the dc capacitor bus in Figure 4.19b.

SIMULATION PROBLEMS

4.15 Simulate a two-quadrant pole of Figure 4.7a in dc steady state. The nominal values are as follows: $V_d = 200$ V, and the output has in series $R_a = 0.37 \, \Omega$, $L_a = 1.5$ mH, and $E_a = 136$ V. $\hat{V}_{tri} = 1$ V. The switching frequency $f_s = 20$ kHz. In dc steady state, the average output current is $I_a = 10$ A. (a) Obtain the plot of $v_{aN}(t)$, $i_a(t)$, and $i_{da}(t)$, (b) obtain the peak-peak ripple in $i_a(t)$ and compare it with its value obtained analytically, and (c) obtain the average values of $i_a(t)$ and $i_{da}(t)$, and show that these two averages are related by the duty-ratio d_a.

4.16 Repeat Problem 4.15 by calculating the value of the control voltage such that the converter pole is operating in the boost mode, with $I_a = -10$ A.

DC-DC CONVERTERS

4.17 Simulate the dc-dc converter of Figure 4.11a in dc steady state. The nominal values are as follows: $V_d = 200$ V, and the output consists of in series $R_a = 0.37 \, \Omega$, $L_a = 1.5$ mH, and $E_a = 136$ V. $\hat{V}_{tri} = 1$ V. The switching frequency $f_s = 20$ kHz. In the dc steady state, the average output current is $I_a = 10$ A.

(a) Obtain the plot of $v_o(t)$, $i_o(t)$, and $i_d(t)$; (b) obtain the peak-peak ripple in $i_o(t)$ and compare it with its value obtained analytically; and (c) obtain the average values of $i_o(t)$ and $i_d(t)$, and show that these two averages are related by the duty-ratio d in Equation 4.17.

4.18   In Problem 4.17, apply a step-increase at 0.5 ms in the control voltage to reach the output current of 15 A (in steady state) and observe the output current response.

4.19   Repeat Problem 4.18 with each converter pole represented on its average basis.

DC-TO-THREE-PHASE AC INVERTERS

4.20   Simulate the three-phase ac inverter on an average basis for the system described in Problem 4.13. Obtain the various waveforms.

4.21   Repeat Problem 4.20 for a corresponding switching circuit and compare the switching waveforms with the average waveforms in Problem 4.20.

5

MAGNETIC CIRCUITS

5.1 INTRODUCTION

The purpose of this chapter is to review some of the basic concepts associated with magnetic circuits and to develop an understanding of transformers, which is needed for the study of ac motors and generators.

5.2 MAGNETIC FIELD PRODUCED BY CURRENT-CARRYING CONDUCTORS

When a current i is passed through a conductor, a magnetic field is produced. The direction of the magnetic field depends on the direction of the current. As shown in Figure 5.1a, the current through a conductor, perpendicular and *into* the paper plane, is represented by "$\times$"; this current produces magnetic field in a clockwise direction. Conversely, the current *out of* the paper plane, represented by a dot, produces magnetic field in a counter-clockwise direction, as shown in Figure 5.1b.

5.2.1 Ampere's Law

The magnetic-field intensity H produced by current-carrying conductors can be obtained by means of Ampere's Law, which in its simplest form states that, at any time, the line (contour) integral of the magnetic field intensity along *any* closed path equals the total current enclosed by this path. Therefore, in Figure 5.1c,

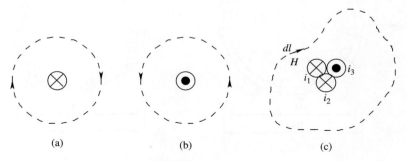

    (a)   (b)   (c)

FIGURE 5.1 Magnetic field; Ampere's Law.

$$\oint Hd\ell = \sum i \tag{5.1}$$

where $\oint$ represents a contour or a closed-line integration. Note that the scalar H in Equation 5.1 is the component of the magnetic field intensity (a vector field) in the direction of the differential length $d\ell$ along the closed path. Alternatively, we can express the field intensity and the differential length to be vector quantities, which will require a dot product on the left side of Equation 5.1.

Example 5.1
Consider the coil in Figure 5.2, which has $N = 25$ turns. The toroid on which the coil is wound has an inside diameter $ID = 5$ cm and an outside diameter $OD = 5.5$ cm. For a current $i = 3$ A, calculate the field intensity H along the mean-path length within the toroid.

Solution Due to symmetry, the magnetic field intensity H_m along a circular contour within the toroid is constant. In Figure 5.2, the mean radius $r_m = \frac{1}{2}(\frac{OD+ID}{2})$. Therefore, the mean path of length $\ell_m (= 2\pi r_m = 0.165\ m)$ encloses the current i N-times, as shown in Figure 5.2b. Therefore, from Ampere's Law in Equation 5.1, the field intensity along this mean path is

$$H_m = \frac{N\,i}{\ell_m} \tag{5.2}$$

which for the given values can be calculated as

$$H_m = \frac{25 \times 3}{0.165} = 454.5 \text{ A/m}.$$

If the width of the toroid is much smaller than the mean radius r_m, it is reasonable to assume a uniform H_m throughout the cross-section of the toroid.

  The field intensity in Equation 5.2 has the units of [A/m], noting that "turns" is a unit-less quantity. The product Ni is commonly referred to as the ampere-turns or mmf F

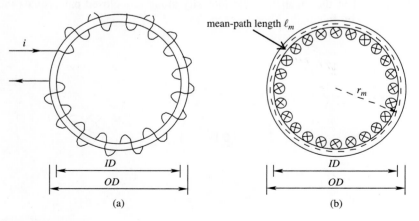

FIGURE 5.2   Toroid.

that produces the magnetic field. The current in Equation 5.2 may be dc, or time varying. If the current is time varying, the relationship in Equation 5.2 is valid on an instantaneous basis; that is, $H_m(t)$ is related to $i(t)$ by N/ℓ_m.

5.3 FLUX DENSITY B AND THE FLUX ϕ

At any instant of time t for a given H-field, the density of flux lines, called the flux density B (in units of $[T]$ for Tesla) depends on the permeability μ of the material on which this H-field is acting. In air,

$$B = \mu_o H \quad \mu_o = 4\pi \times 10^{-7} \left[\frac{henries}{m} \right] \tag{5.3}$$

where μ_o is the permeability of air or free space.

5.3.1 Ferromagnetic Materials

Ferromagnetic materials guide magnetic fields and, due to their high permeability, require small ampere-turns (a small current for a given number of turns) to produce the desired flux density. These materials exhibit the multi-valued nonlinear behavior shown by their B-H characteristics in Figure 5.3a. Imagine that the toroid in Figure 5.2 consists of a ferromagnetic material such as silicon steel. If the current through the coil is slowly varied in a sinusoidal manner with time, the corresponding H-field will cause one of the hysteresis loops shown in Figure 5.3a to be traced. Completing the loop once results in a net dissipation of energy within the material, causing power loss referred as the hysteresis loss.

Increasing the peak value of the sinusoidally-varying H-field will result in a bigger hysteresis loop. Joining the peaks of the hysteresis loop, we can approximate the B-H characteristic by the single curve shown in Figure 5.3b. At low values of magnetic field, the B-H characteristic is assumed to be linear with a constant slope, such that

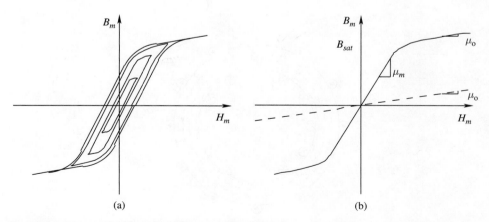

(a)   (b)

FIGURE 5.3 *B-H* Characteristics of ferromagnetic materials.

$$B_m = \mu_m H_m \qquad (5.4a)$$

where μ_m is the permeability of the ferromagnetic material. Typically, the μ_m of a material is expressed in terms of a permeability μ_r relative to the permeability of air:

$$\mu_m = \mu_r \mu_o \qquad \left(\mu_r = \frac{\mu_m}{\mu_o} \right) \qquad (5.4b)$$

In ferromagnetic materials, the μ_m can be several thousand times larger than the μ_o.

In Figure 5.3b, the linear relationship (with a constant μ_m) is approximately valid until the "knee" of the curve is reached, beyond which the material begins to saturate. Ferromagnetic materials are often operated up to a maximum flux density, slightly above the "knee" of 1.6 T to 1.8 T, beyond which many more ampere-turns are required to increase flux density only slightly. In the saturated region, the incremental permeability of the magnetic material approaches μ_o, as shown by the slope of the curve in Figure 5.3b.

In this course, we will assume that the magnetic material is operating in its linear region and therefore its characteristic can be represented by $B_m = \mu_m H_m$, where μ_m remains constant.

5.3.2   Flux ϕ

Magnetic flux lines form closed paths, as shown in Figure 5.4's toroidal magnetic core, which is surrounded by the current-carrying coil. The flux in the toroid can be calculated by selecting a circular area A_m in a plane perpendicular to the direction of the flux lines. As discussed in Example 5.1, it is reasonable to assume a uniform H_m and hence a uniform flux-density B_m throughout the core cross-section.

Substituting for H_m from Equation 5.2 into Equation 5.4a,

$$B_m = \mu_m \frac{Ni}{\ell_m} \qquad (5.5)$$

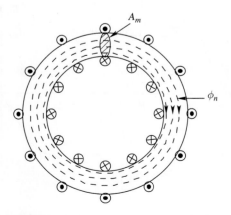

FIGURE 5.4   Toroid with flux ϕ_m.

where B_m is the density of flux lines in the core. Therefore, making the assumption of a uniform B_m, the flux ϕ_m can be calculated as

$$\phi_m = B_m A_m \tag{5.6}$$

where flux has the units of Weber [Wb]. Substituting for B_m from Equation 5.5 into Equation 5.6,

$$\phi_m = A_m \left(\mu_m \frac{Ni}{\ell_m} \right) = \frac{Ni}{\underbrace{\left(\frac{\ell_m}{\mu_m A_m} \right)}_{\mathfrak{R}_m}} \tag{5.7}$$

where Ni equals the ampere-turns (or mmf F) applied to the core, and the term within the brackets on the right side is called the reluctance $\mathfrak{R}_m$ of the magnetic core. From Equation 5.7

$$\mathfrak{R}_m = \frac{\ell_m}{\mu_m A_m} \quad [A/Wb] \tag{5.8}$$

Equation 5.7 makes it clear that the reluctance has the units [A/Wb]. Equation 5.8 shows that the reluctance of a magnetic structure, for example the toroid in Figure 5.4, is linearly proportional to its magnetic path length and inversely proportional to both its cross-sectional area and the permeability of its material.

Equation 5.7 shows that the amount of flux produced by the applied ampere-turns $F (= Ni)$ is inversely proportional to the reluctance $\mathfrak{R}$; this relationship is analogous to Ohm's Law $(I = V/R)$ in electric circuits in dc steady state.

5.3.3 Flux Linkage

If all turns of a coil, for example the one in Figure 5.4, are linked by the same flux ϕ, then the coil has a flux linkage λ, where

$$\lambda = N\phi \tag{5.9}$$

Example 5.2
In Example 5.1, the core consists of a material with $\mu_r = 2000$. Calculate the flux density B_m and the flux ϕ_m.

Solution   In Example 5.1, we calculated that $H_m = 454.5 \ A/m$. Using Equations 5.4a and 5.4b, $B_m = 4\pi \times 10^{-7} \times 2000 \times 454.5 = 1.14$ T. The toroid width is $\frac{OD-ID}{2} = 0.25 \times 10^{-2}$ m. Therefore, the cross-sectional area of the toroid is

$$A_m = \frac{\pi}{4}(0.25 \times 10^{-2})^2 = 4.9 \times 10^{-6} \ \text{m}^2$$

Hence, from Equation 5.6, assuming that the flux density is uniform throughout the cross-section,

$$\phi_m = 1.14 \times 4.9 \times 10^{-6} = 5.59 \times 10^{-6} \ \text{Wb}$$

5.4 MAGNETIC STRUCTURES WITH AIR GAPS

In the magnetic structures of electric machines, the flux lines have to cross two air gaps. To study the effect of air gaps, let us consider the simple magnetic structure of Figure 5.5 consisting of an N-turn coil on a magnetic core made up of iron. The objective is to establish a desired magnetic field in the air gap of length ℓ_g by controlling the coil current i. We will assume the magnetic field intensity H_m to be uniform along the mean path length ℓ_m in the magnetic core. The magnetic field intensity in the air gap is denoted as H_g. From Ampere's Law in Equation 5.1, the line integral along the mean path within the core and in the air gap yields the following equation:

$$H_m\ell_m + H_g\ell_g = Ni \tag{5.10}$$

Applying Equation 5.3 to the air gap and Equation 5.4a to the core, the flux densities corresponding to H_m and H_g are

$$B_m = \mu_m H_m \quad \text{and} \quad B_g = \mu_o H_g \tag{5.11}$$

In terms of the flux densities of Equation 5.11, Equation 5.10 can be written as

$$\frac{B_m}{\mu_m}\ell_m + \frac{B_g}{\mu_o}\ell_g = Ni \tag{5.12}$$

Since flux lines form closed paths, the flux crossing any perpendicular cross-sectional area in the core is the same as that crossing the air gap (neglecting the leakage flux, which is discussed later on). Therefore,

$$\phi = A_m B_m = A_g B_g \quad \text{or} \tag{5.13}$$

$$B_m = \frac{\phi}{A_m} \quad \text{and} \quad B_g = \frac{\phi}{A_g} \tag{5.14}$$

Generally, flux lines bulge slightly around the air gap, as shown in Figure 5.5. This bulging is called the *fringing effect,* which can be accounted for by estimating the air gap area A_g, which is done by increasing each dimension in Figure 5.5 by the length of the air gap:

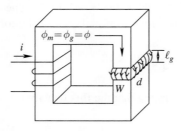

FIGURE 5.5 Magnetic structure with air gap.

$$A_g = (W + \ell_g)(d + \ell_g) \tag{5.15}$$

Substituting flux densities from Equation 5.14 into Equation 5.12,

$$\phi \left(\frac{\ell_m}{A_m \mu_m} + \frac{\ell_g}{A_g \mu_o} \right) = Ni \tag{5.16}$$

In Equation 5.16, we can recognize from Equation 5.8 that the two terms within the parenthesis equal the reluctances of the core and of the air gap, respectively. Therefore, the effective reluctance $\mathfrak{R}$ of the whole structure in the path of the flux lines is the sum of the two reluctances:

$$\mathfrak{R} = \mathfrak{R}_m + \mathfrak{R}_g \tag{5.17}$$

Substituting from Equation 5.17 into Equation 5.16, where Ni equals the applied mmf F,

$$\phi = \frac{F}{\mathfrak{R}} \tag{5.18}$$

Equation 5.18 allows ϕ to be calculated for the applied ampere-turns (mmf F). Then B_m and B_g can be calculated from Equation 5.14.

Example 5.3

In the structure of Figure 5.5, all flux lines in the core are assumed to cross the air gap. The structure dimensions are as follows: core cross-sectional area $A_m = 20$ cm^2, mean path length $\ell_m = 40$ cm, $\ell_g = 2$ mm, and $N = 75$ turns. In the linear region, the core permeability can be assumed to be constant, with $\mu_r = 4500$. The coil current i ($= 30$ A) is below the saturation level. Ignore the flux fringing effect. Calculate the flux density in the air gap, (a) including the reluctance of the core as well as that of the air gap and (b) ignoring the core reluctance in comparison to the reluctance of the air gap.

Solution   From Equation 5.8,

$$\mathfrak{R}_m = \frac{\ell_m}{\mu_o \mu_r A_m} = \frac{40 \times 10^{-2}}{4\pi \times 10^{-7} \times 4500 \times 20 \times 10^{-4}} = 3.54 \times 10^4 \, \frac{A}{Wb}, \text{ and}$$

$$\mathfrak{R}_g = \frac{\ell_g}{\mu_o A_g} = \frac{2 \times 10^{-3}}{4\pi \times 10^{-7} \times 20 \times 10^{-4}} = 79.57 \times 10^4 \, \frac{A}{Wb}$$

a. Including both reluctances, from Equation 5.16,

$$\phi_g = \frac{Ni}{\mathfrak{R}_m + \mathfrak{R}_g} \quad \text{and}$$

$$B_g = \frac{\phi_g}{A_g} = \frac{Ni}{(\mathfrak{R}_m + \mathfrak{R}_g)A_g} = \frac{75 \times 30}{(79.57 + 3.54) \times 10^4 \times 20 \times 10^{-4}} = 1.35 \text{ T}$$

b. Ignoring the core reluctance, from Equation 5.16,

$$\phi_g = \frac{Ni}{\mathfrak{R}_g} \quad \text{and}$$

$$B_g = \frac{\phi_g}{A_g} = \frac{Ni}{\mathfrak{R}_g A_g} = \frac{75 \times 30}{79.57 \times 10^4 \times 20 \times 10^{-4}} = 1.41 \text{ T}$$

This example shows that the reluctance of the air gap dominates the flux and the flux density calculations; thus we can often ignore the reluctance of the core in comparison to that of the air gap.

5.5 INDUCTANCES

At any instant of time in the coil of Figure 5.6a, the flux linkage of the coil (due to flux lines entirely in the core) is related to the current i by a parameter defined as the inductance L_m:

$$\lambda_m = L_m i \qquad (5.19)$$

where the inductance $L_m (= \lambda_m/i)$ is constant if the core material is in its linear operating region.

The coil inductance in the linear magnetic region can be calculated by multiplying all the factors shown in Figure 5.6b, which are based on earlier equations:

$$L_m = \underbrace{\left(\frac{N}{\ell_m}\right)}_{\text{Eq.5.2}} \underbrace{\mu_m}_{\text{Eq.5.4a}} \underbrace{A_m}_{\text{Eq.5.6}} \underbrace{N}_{\text{Eq.5.9}} = \frac{N^2}{\left(\frac{\ell_m}{\mu_m A_m}\right)} = \frac{N^2}{\mathfrak{R}_m} \qquad (5.20)$$

Equation 5.20 indicates that the inductance L_m is strictly a property of the magnetic circuit (i.e., the core material, the geometry, and the number of turns), provided that the operation is in the linear range of the magnetic material, where the slope of its *B-H* characteristic can be represented by a constant μ_m.

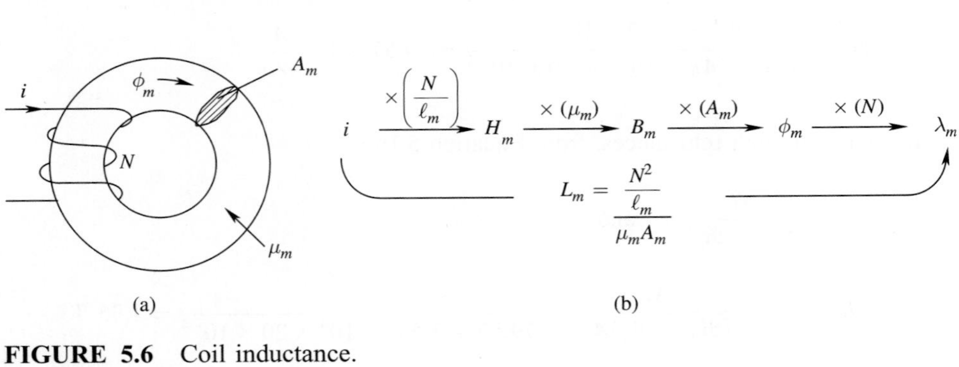

FIGURE 5.6 Coil inductance.

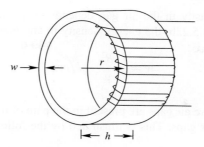

FIGURE 5.7 Rectangular toroid.

Example 5.4

In the rectangular toroid of Figure 5.7, $w = 5$ mm, $h = 15$ mm, the mean path length $\ell_m = 18$ cm, $\mu_r = 5000$, and $N = 100$ turns. Calculate the coil inductance L_m, assuming that the core is unsaturated.

Solution   From Equation 5.8,

$$\mathfrak{R}_m = \frac{\ell_m}{\mu_m A_m} = \frac{0.18}{5000 \times 4\pi \times 10^{-7} \times 5 \times 10^{-3} \times 15 \times 10^{-3}} = 38.2 \times 10^4 \frac{A}{Wb}$$

Therefore, from Equation 5.20,

$$L_m = \frac{N^2}{\mathfrak{R}_m} = 26.18 \ mH$$

5.5.1 Magnetic Energy Storage in Inductors

Energy in an inductor is stored in its magnetic field. From the study of electric circuits, we know that at any time, with a current i, the energy stored in the inductor is

$$W = \frac{1}{2} L_m \, i^2 \ [J] \tag{5.21}$$

where [J], for Joules, is a unit of energy. Initially assuming a structure without an air gap, such as in Figure 5.6a, we can express the energy storage in terms of flux density, by substituting into Equation 5.21 the inductance from Equation 5.20 and the current from the Ampere's Law in Equation 5.2:

$$W_m = \frac{1}{2} \underbrace{\frac{N^2}{\frac{\ell_m}{\mu_m A_m}}}_{} \underbrace{(H_m \ell_m / N)^2}_{i^2} = \frac{1}{2} \frac{(H_m \ell_m)^2}{\frac{\ell_m}{\mu_m A_m}} = \frac{1}{2} \frac{B_m^2}{\mu_m} \underbrace{A_m \, \ell_m}_{volume} \ [J] \tag{5.22a}$$

where $A_m \, \ell_m = volume$, and in the linear region $B_m = \mu_m H_m$. Therefore, from Equation 5.22a, the energy density in the core is

$$w_m = \frac{1}{2} \frac{B_m^2}{\mu_m} \tag{5.22b}$$

Similarly, the energy density in the air gap depends on μ_o and the flux density in it. Therefore, from Equation 5.22b, the energy density in any medium can be expressed as

$$w = \frac{1}{2}\frac{B^2}{\mu} \quad [J/m^3] \tag{5.23}$$

In electric machines, where air gaps are present in the path of the flux lines, the energy is primarily stored in the air gaps. This is illustrated by the following example.

Example 5.5
In Example 5.3 part (a), calculate the energy stored in the core and in the air gap and compare the two.

Solution In Example 5.3 part (a), $B_m = B_g = 1.35$ T. Therefore, from Equation 5.23,

$$w_m = \frac{1}{2}\frac{B_m{}^2}{\mu_m} = 161.1 \ J/m^3 \quad \text{and}$$

$$w_g = \frac{1}{2}\frac{B_g{}^2}{\mu_o} = 0.725 \times 10^6 \ J/m^3.$$

Therefore, $\dfrac{w_g}{w_m} = \mu_r = 4500$.

Based on the given cross-sectional areas and lengths, the core volume is 200 times larger than that of the air gap. Therefore, the ratio of the energy storage is

$$\frac{W_g}{W_m} = \frac{w_g}{w_m} \times \frac{(volume)_g}{(volume)_m} = \frac{4500}{200} = 22.5$$

5.6 FARADAY'S LAW: INDUCED VOLTAGE IN A COIL DUE TO TIME-RATE OF CHANGE OF FLUX LINKAGE

In our discussion so far, we have established in magnetic circuits relationships between the electrical quantity i and the magnetic quantities H, B, ϕ, and λ. These relationships are valid under dc (static) conditions, as well as at any instant when these quantities are varying with time. We will now examine the voltage across the coil under time-*varying* conditions. In the coil of Figure 5.8, Faraday's Law dictates that the time-rate of change of flux-linkage equals the voltage across the coil at any instant:

$$e(t) = \frac{d}{dt}\lambda(t) = N\frac{d}{dt}\phi(t) \tag{5.24}$$

This assumes that all flux lines link all N-turns such that $\lambda = N\phi$. The polarity of the emf $e(t)$ and the direction of $\phi(t)$ in the above equation are yet to be justified.

The above relationship is valid, no matter what is causing the flux to change. One possibility is that a second coil is placed on the same core. When the second coil is

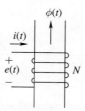

FIGURE 5.8 Voltage polarity and direction of flux and current.

supplied by a time-varying current, mutual coupling causes the flux ϕ through the coil shown in Figure 5.8 to change with time. The other possibility is that a voltage $e(t)$ is applied across the coil in Figure 5.8, causing the change in flux, which can be calculated by integrating both sides of Equation 5.24 with respect to time:

$$\phi(t) = \phi(0) + \frac{1}{N} \int_0^t e(\tau) \cdot d\tau \tag{5.25}$$

where $\phi(0)$ is the initial flux at $t = 0$ and τ is a variable of integration.

Recalling the Ohm's Law equation $v = Ri$, the current direction through a resistor is defined to be into the terminal chosen to be of the positive polarity. This is the passive sign convention. Similarly, in the coil of Figure 5.8, we can establish the voltage polarity and the flux direction in order to apply Faraday's Law, given by Equations 5.24 and 5.25. If the flux direction is given, we can establish the voltage polarity as follows: first determine the direction of a hypothetical current that will produce flux in the same direction as given. Then, the positive polarity for the voltage is at the terminal that this hypothetical current is entering. Conversely, if the voltage polarity is given, imagine a hypothetical current entering the positive-polarity terminal. This current, based on how the coil is wound, for example in Figure 5.8, determines the flux direction for use in Equations 5.24 and 5.25.

Another way to determine the polarity of the induced emf is to apply Lenz's Law, which states the following: if a current is allowed to flow due to the voltage induced by an increasing flux-linkage, for example, the direction of this hypothetical current will be to oppose the flux change.

Example 5.6

In the structure of Figure 5.8, the flux $\phi_m (= \hat{\phi}_m \sin \omega t)$ linking the coil is varying sinusoidally with time, where $N = 300$ turns, $f = 60$ Hz, and the cross-sectional area $A_m = 10$ cm^2. The peak flux density $\hat{B}_m = 1.5$ T. Calculate the expression for the induced voltage with the polarity shown in Figure 5.8. Plot the flux and the induced voltage as functions of time.

Solution From Equation 5.6, $\hat{\phi}_m = \hat{B}_m A_m = 1.5 \times 10 \times 10^{-4} = 1.5 \times 10^{-3}$ Wb. From Faraday's Law in Equation 5.24, $e(t) = \omega N \hat{\phi}_m \cos \omega t = 2\pi \times 60 \times 300 \times 1.5 \times 10^{-3} \times \cos \omega t = 169.65 \cos \omega t$ V. The waveforms are plotted in Figure 5.9.

Example 5.6 illustrates that the voltage is induced due to $d\phi/dt$, regardless of whether any current flows in that coil. In the following subsection, we will establish the relationship between $e(t)$, $\phi(t)$, and $i(t)$.

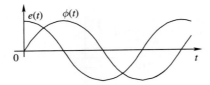

FIGURE 5.9   Waveforms of flux and induced voltage.

5.6.1   Relating $e(t)$, $\phi(t)$, and $i(t)$

In the coil of Figure 5.10a, an applied voltage $e(t)$ results in $\phi(t)$, which is dictated by the Faraday's Law equation in the integral form, Equation 5.25. But what about the current drawn by the coil to establish this flux? Rather than going back to Ampere's Law, we can express the coil flux linkage in terms of its inductance and current using Equation 5.19:

$$\lambda(t) = Li(t) \tag{5.26}$$

Assuming that the entire flux links all N turns, the coil flux linkage $\lambda(t) = N\phi(t)$. Substituting this into Equation 5.26 gives

$$\phi(t) = \frac{L}{N}i(t) \tag{5.27}$$

Substituting for $\phi(t)$ from Equation 5.27 into Faraday's Law in Equation 5.24 results in

$$e(t) = N\frac{d\phi}{dt} = L\frac{di}{dt} \tag{5.28}$$

Equations 5.27 and 5.28 relate $i(t)$, $\phi(t)$, and $e(t)$; all of these are plotted in Figure 5.10b.

Example 5.7
In Example 5.6, the coil inductance is 50 mH. Calculate the expression for the current $i(t)$ in Figure 5.10b.

Solution   From Equation 5.27, $i(t) = \frac{N}{L}\phi(t) = \frac{300}{50 \times 10^{-3}} 1.5 \times 10^{-3} \sin \omega t = 9.0 \sin \omega t$ A

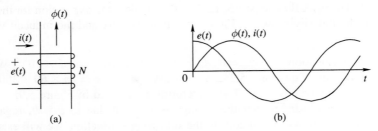

(a)   (b)

FIGURE 5.10   Voltage, current, and flux.

5.7 LEAKAGE AND MAGNETIZING INDUCTANCES

Just as conductors guide currents in electric circuits, magnetic cores guide *flux* in *magnetic circuits*. But there is an important difference. In electric circuits, the conductivity of copper is approximately 10^{20} times higher than that of air, allowing leakage currents to be neglected at dc or at low frequencies such as 60 Hz. In magnetic circuits, however, the permeabilities of magnetic materials are only around 10^4 times greater than that of air. Because of this relatively low ratio, the core window in the structure of Figure 5.11a has "leakage" flux lines, which do not reach their intended destination—the air gap. Note that the coil shown in Figure 5.11a is drawn schematically. In practice, the coil consists of multiple layers, and the core is designed to fit as snugly to the coil as possible, thus minimizing the unused "window" area.

The leakage effect makes accurate analysis of magnetic circuits more difficult, so that it requires sophisticated numerical methods, such as finite element analysis. However, we can account for the effect of leakage fluxes by making certain approximations. We can divide the total flux ϕ into two parts: the magnetic flux ϕ_m, which is completely confined to the core and links all N turns, and the leakage flux, which is partially or entirely in air and is represented by an "equivalent" leakage flux ϕ_ℓ, which also links all N turns of the coil but does not follow the entire magnetic path, as shown in Figure 5.11b. Thus,

$$\phi = \phi_m + \phi_\ell \tag{5.29}$$

where ϕ is the equivalent flux which links all N turns. Therefore, the total flux linkage of the coil is

$$\lambda = N\phi = \underbrace{N\phi_m}_{\lambda_m} + \underbrace{N\phi_\ell}_{\lambda_\ell} = \lambda_m + \lambda_\ell \tag{5.30}$$

The total inductance (called the self-inductance) can be obtained by dividing both sides of Equation 5.30 by the current i:

$$\underbrace{\frac{\lambda}{i}}_{L_{self}} = \underbrace{\frac{\lambda_m}{i}}_{L_m} + \underbrace{\frac{\lambda_\ell}{i}}_{L_\ell} \tag{5.31}$$

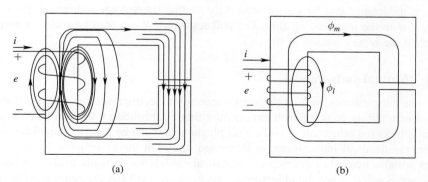

(a)　　　　　(b)

FIGURE 5.11 (a) Magnetic and leakage fluxes; (b) Equivalent representation of these fluxes.

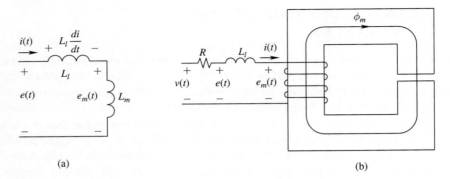

(a)   (b)

FIGURE 5.12   (a) Circuit representation; (b) Leakage inductance separated from the core.

Therefore,

$$L_{self} = L_m + L_\ell \tag{5.32}$$

where L_m is often called the *magnetizing inductance* due to ϕ_m in the magnetic core, and L_ℓ is called the *leakage inductance* due to the leakage flux ϕ_ℓ. From Equation 5.32, the total flux linkage of the coil can be written as

$$\lambda = (L_m + L_\ell)i \tag{5.33}$$

Hence, from Faraday's Law in Equation 5.24,

$$e(t) = L_\ell \frac{di}{dt} + \underbrace{L_m \frac{di}{dt}}_{e_m(t)} \tag{5.34}$$

This results in the circuit of Figure 5.12a. In Figure 5.12b, the voltage drop due to the leakage inductance can be shown separately so that the voltage induced in the coil is solely due to the magnetizing flux. The coil resistance R can then be added in series to complete the representation of the coil.

5.7.1   Mutual Inductances

Most magnetic circuits, such as those encountered in electric machines and transformers, consist of multiple coils. In such circuits, the flux established by the current in one coil partially links the other coil or coils. This phenomenon can be described mathematically by means of mutual inductances, as examined in circuit theory courses. Mutual inductances are also needed to develop mathematical models for dynamic analysis of electric machines. Since it is not the objective of this book, we will not elaborate any further on the topic of mutual inductances. Rather, we will use simpler and more intuitive means to accomplish the task at hand.

5.8 TRANSFORMERS

Electric machines consist of several mutually-coupled coils where a portion of the flux produced by one coil (winding) links other windings. A transformer consists of two or more tightly-coupled windings where almost all of the flux produced by one winding links the other windings. Transformers are essential for transmission and distribution of electric power. They also facilitate the understanding of ac motors and generators very effectively.

To understand the operating principles of transformers, consider a single coil, also called a winding of N_1 turns, as shown in Figure 5.13a. Initially, we will assume that the resistance and the leakage inductance of this winding are both zero; the second assumption implies that all of the flux produced by this winding is confined to the core Applying a time-varying voltage e_1 to this winding results in a flux $\phi_m(t)$. From Faraday's Law:

$$e_1(t) = N_1 \frac{d\phi_m}{dt} \tag{5.35}$$

where $\phi_m(t)$ is completely dictated by the time-integral of the applied voltage, as given below (where it is assumed that the flux in the winding is initially zero):

$$\phi_m(t) = \frac{1}{N_1} \int_0^t e_1(\tau) \cdot d\tau \tag{5.36}$$

The current $i_m(t)$ drawn to establish this flux depends on the magnetizing inductance L_m of this winding, as depicted in Figure 5.13b.

A second winding of N_2 turns is now placed on the core, as shown in Figure 5.13a. A voltage is induced in the second winding due to the flux $\phi_m(t)$ linking it. From Faraday's Law,

$$e_2(t) = N_2 \frac{d\phi_m}{dt} \tag{5.37}$$

Equations 5.35 and 5.37 show that in each winding, the volts-per-turn are the same, due to the same $d\phi_m/dt$:

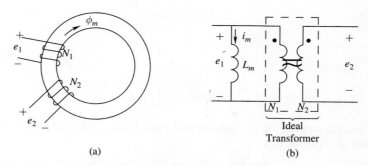

(a)   (b)

FIGURE 5.13 (a) Core with two coils; (b) Equivalent circuit.

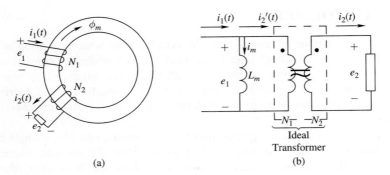

FIGURE 5.14   (a) Transformer with impedance on the secondary; (b) Equivalent circuit.

$$\frac{e_1(t)}{N_1} = \frac{e_2(t)}{N_2} \tag{5.38}$$

We can represent the relationship of Equation 5.38 in Figure 5.14b by means of a hypothetical circuit component called the "ideal transformer," which relates the voltages in the two windings by the turns-ratio N_1/N_2:

$$\frac{e_1(t)}{e_2(t)} = \frac{N_1}{N_2} \tag{5.39}$$

The dots in Figure 5.14b convey the information that the winding voltages will be of the same polarity at the dotted terminals with respect to their undotted terminals. For example, if ϕ_m is increasing with time, the voltages at both dotted terminals will be positive with respect to the corresponding undotted terminals. The advantage of using this dot convention is that the winding orientations on the core need not be shown in detail.

A load such an *R-L* combination is now connected across the secondary winding, as shown in Figure 5.14a. A current $i_2(t)$ will now flow through the *R-L* combination. The resulting ampere-turns $N_2 i_2$ will tend to change the core flux ϕ_m but *cannot* because $\phi_m(t)$ is completely dictated by the applied voltage $e_1(t)$, as given in Equation 5.36. Therefore, additional current i_2' in Figure 5.14b is drawn by winding 1 in order to compensate (or nullify) $N_2 i_2$, such that

$$N_1 i_2' = N_2 i_2 \tag{5.40}$$

or

$$\frac{i_2'(t)}{i_2(t)} = \frac{N_2}{N_1} \tag{5.41}$$

This is the second property of the "ideal transformer." Thus, the total current drawn from the terminals of winding 1 is

$$i_1(t) = i_m(t) + i_2'(t) \tag{5.42}$$

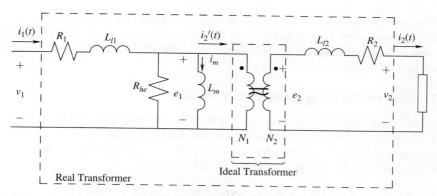

FIGURE 5.15 Equivalent circuit of a real transformer.

In Figure 5.14, the resistance and the leakage inductance associated with winding 2 appear in series with the *R-L* load. Therefore, the induced voltage e_2 differs from the voltage v_2 at the winding terminals by the voltage drop across the winding resistance and the leakage inductance, as depicted in Figure 5.15. Similarly, the applied voltage v_1 differs from the emf e_1 (induced by the time-rate of change of the flux ϕ_m) by the voltage drop across the resistance and the leakage inductance of winding 1.

5.8.1 Core Losses

We can model core losses due to hysteresis and eddy currents by connecting a resistance R_{he} in parallel with L_m, as shown in Figure 5.15. The loss due to the hysteresis-loop in the *B-H* characteristic was discussed earlier. Another source of core loss is due to eddy currents. All magnetic materials have a finite electrical resistivity (ideally, it should be infinite). As discussed in section 5.6, which dealt with Faraday's voltage induction law, time-varying fluxes induce voltages in the core, which result in circulating (eddy) currents within the core to oppose these flux changes (and partially neutralize them).

In Figure 5.16a, an increasing flux ϕ will set up many current loops (due to induced voltages that oppose the change in core flux), which result in losses. The primary means of limiting the eddy-current losses is to make the core out of steel laminations that are insulated from each other by means of thin layers of varnish, as shown in Figure 5.16b. A few laminations are shown to illustrate how insulated laminations reduce eddy-current losses. Because of the insulation between laminations, the current is forced to flow in

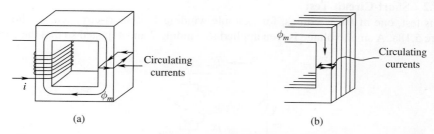

(a)   (b)

FIGURE 5.16 (a) Eddy-currents induced by time-varying fluxes; (b) Core with insulated laminations.

much smaller loops within each lamination. Laminating the core reduces the flux and the induced voltage more than it reduces the effective resistance to the currents within a lamination, thus reducing the overall losses. For 50- or 60-Hz operation, lamination thicknesses are typically 0.2 to 1 mm.

5.8.2 Real versus Ideal Transformer Models

Consider the equivalent circuit of a real transformer, shown in Figure 5.15. If we ignore all of the parasitics, such as the leakage inductances and the losses, and if we assume that the core permeability is infinite (thus $L_m = \infty$), then the equivalent circuit of the real transformer reduces to exactly that of the ideal transformer.

5.8.3 Determining Transformer Model Parameters

In order to utilize the transformer equivalent circuit of Figure 5.15, we need to determine the values of its various parameters. We can obtain these by means of two tests: (1) an open-circuit test and (2) a short-circuit test.

5.8.3.1 Open-Circuit Test

In this test, one of the windings, for example winding 2, is kept open as shown in Figure 5.17, while winding 1 is applied its rated voltage. The rms input voltage V_{oc}, the rms current I_{oc} and the average power P_{oc} are measured, where the subscript "*oc*" refers to the open-circuit condition. Under the open-circuit condition, the winding current is very small and is determined by the large magnetizing impedance. Therefore, the voltage drop across the leakage impedance can be neglected, as shown in Figure 5.17. In terms of the measured quantities, R_{he} can be calculated as follows:

$$R_{he} = \frac{V_{oc}^{2}}{P_{oc}} \tag{5.43}$$

The magnitude of the open-circuit impedance in Figure 5.17 can be calculated as

$$|Z_{oc}| = \frac{X_m R_{he}}{\sqrt{R_{he}^{2} + X_m^{2}}} = \frac{V_{oc}}{I_{oc}} \tag{5.44}$$

Using the measured values V_{oc}, I_{oc}, and R_{he} calculated from Equation 5.43, we can calculate the magnetizing reactance X_m from Equation 5.44.

5.8.3.2 Short-Circuit Test

In this test, one of the windings, for example winding 1, is short-circuited, as shown in Figure 5.18a. A small voltage is then applied to winding 2 and adjusted so that the current

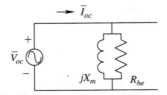

FIGURE 5.17 Open-circuit test.

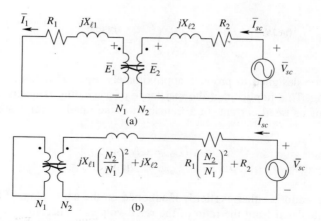

FIGURE 5.18 Short-circuit test.

in each winding is approximately equal to its rated value. Under this short-circuited condition, the magnetizing reactance X_m and the core-loss resistance R_{he} can be neglected in comparison to the leakage impedance of winding 1, as shown in Figure 5.18a.

In this circuit, the rms voltage V_{sc}, the rms current I_{sc}, and the average power P_{sc} are measured, where the subscript "sc" represents the short-circuited condition.

In terms of voltages, currents, and the turns-ratio defined in Figure 5.18a,

$$\frac{\overline{E}_2}{\overline{E}_1} = \frac{N_2}{N_1} \quad \text{and} \quad \frac{\overline{I}_{sc}}{\overline{I}_1} = \frac{N_1}{N_2} \tag{5.45}$$

Therefore,

$$\frac{\overline{E}_2}{\overline{I}_{sc}} = \frac{\frac{N_2}{N_1}\overline{E}_1}{\frac{N_1}{N_2}\overline{I}_1} = \left(\frac{N_2}{N_1}\right)^2 \frac{\overline{E}_1}{\overline{I}_1} \tag{5.46}$$

Notice that in Equation 5.46, from Figure 5.18a

$$\frac{\overline{E}_1}{\overline{I}_1} = R_1 + jX_{\ell 1} \tag{5.47}$$

Therefore, substituting Equation 5.47 into Equation 5.46,

$$\frac{\overline{E}_2}{\overline{I}_{sc}} = \left(\frac{N_2}{N_1}\right)^2 (R_1 + jX_{\ell 1}) \tag{5.48}$$

This allows the equivalent circuit under the short-circuited condition in Figure 5.18a to be drawn as in Figure 5.18b, where the parasitic components of coil 1 have been moved to coil 2's side and included with coil 2's parasitic components. Having transferred the winding 1 leakage impedance to the side of winding 2, we can effectively replace the ideal transformer portion of Figure 5.18b with a short. Thus, in terms of the measured quantities,

$$R_2 + \left(\frac{N_2}{N_1}\right)^2 R_1 = \frac{P_{sc}}{I_{sc}^2} \tag{5.49}$$

Transformers are designed to produce approximately equal $I^2 R$ losses (copper losses) in each winding. This implies that the resistance of a winding is inversely proportional to the square of its rated current. In a transformer, the rated currents are related to the turns-ratio as

$$\frac{I_{1,rated}}{I_{2,rated}} = \frac{N_2}{N_1} \tag{5.50}$$

where the turns-ratio is either explicitly mentioned on the name-plate of the transformer or it can be calculated from the ratio of the rated voltages. Therefore,

$$\frac{R_1}{R_2} = \left(\frac{I_{2,rated}}{I_{1,rated}}\right)^2 = \left(\frac{N_1}{N_2}\right)^2 \quad \text{or}$$

$$R_1 \left(\frac{N_2}{N_1}\right)^2 = R_2 \tag{5.51}$$

Substituting Equation 5.51 into Equation 5.49,

$$R_2 = \frac{1}{2}\frac{P_{sc}}{I_{sc}^2} \tag{5.52}$$

and R_1 can be calculated from Equation 5.51.

The leakage reactance in a winding is approximately proportional to the square of its number of turns. Therefore,

$$\frac{X_{\ell 1}}{X_{\ell 2}} = \left(\frac{N_1}{N_2}\right)^2 \quad \text{or} \quad \left(\frac{N_2}{N_1}\right)^2 X_{\ell 1} = X_{\ell 2} \tag{5.53}$$

Using Equations 5.51 and 5.53 in Figure 5.18b,

$$|Z_{sc}| = \sqrt{(2R_2)^2 + (2X_{\ell 2})^2} = \frac{V_{sc}}{I_{sc}} \tag{5.54}$$

Using the measured values V_{sc} and I_{sc}, and R_2 calculated from Equation 5.52, we can calculate $X_{\ell 2}$ from Equation 5.54 and $X_{\ell 1}$ from Equation 5.53.

5.9 PERMANENT MAGNETS

Many electric machines other than induction motors consist of permanent magnets in smaller ratings. However, the use of permanent magnets will undoubtedly extend to machines of higher power ratings, because permanent magnets provide a "free" source of

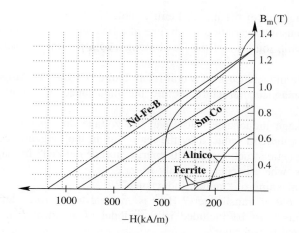

FIGURE 5.19 Characteristics of various permanent-magnet materials.

flux, which otherwise has to be created by current-carrying windings that incur i^2R losses in winding resistance. The higher efficiency and the higher power density offered by permanent-magnet machines are very attractive. In recent years, significant advances have been made in Nd-Fe-B material, which has a very attractive magnetic characteristic, shown in comparison to other permanent-magnet materials in Figure 5.19.

Nd-Fe-B magnets offer high flux density operation, high energy densities, and a high ability to resist demagnetization. Decreases in their manufacturing cost, coupled with advances in operation at high temperatures, will allow their application at much higher power ratings than at present.

In the subsequent chapters, as we discuss permanent-magnet machines, it is adequate for us to treat them as a source of flux, which otherwise would have to be established by current-carrying windings.

SUMMARY/REVIEW QUESTIONS

1. What is the role of magnetic circuits? Why are magnetic materials with very high permeabilities desirable? What is the permeability of air? What is the typical range of the relative permeabilities of ferromagnetic materials like iron?
2. Why can "leakage" be ignored in electric circuits but not in magnetic circuits?
3. What is Ampere's Law, and what quantity is usually calculated by using it?
4. What is the definition of the mmf F?
5. What is meant by "magnetic saturation"?
6. What is the relationship between ϕ and B?
7. How can magnetic reluctance $\Re$ be calculated? What field quantity is calculated by dividing the mmf F by the reluctance $\Re$?
8. In magnetic circuits with an air gap, what usually dominates the total reluctance in the flux path: the air gap or the rest of the magnetic structure?
9. What is the meaning of the flux linkage λ of a coil?
10. Which law allows us to calculate the induced emf? What is the relationship between the induced voltage and the flux linkage?

11. How is the polarity of the induced emf established?
12. Assuming sinusoidal variations with time at a frequency f, how are the rms value of the induced emf, the peak of the flux linking a coil, and the frequency of variation f related?
13. How does the inductance L of a coil relate Faraday's Law to Ampere's Law?
14. In a linear magnetic structure, define the inductance of a coil in terms of its geometry.
15. What is leakage inductance? How can the voltage drop across it be represented separate from the emf induced by the main flux in the magnetic core?
16. In linear magnetic structures, how is energy storage defined? In magnetic structures with air gaps, where is energy mainly stored?
17. What is the meaning of "mutual inductance"?
18. What is the role of transformers? How is an ideal transformer defined? What parasitic elements must be included in the model of an ideal transformer for it to represent a real transformer?
19. What are the advantages of using permanent magnets?

REFERENCES

1. G. R. Slemon, *Electric Machines and Drives* (Addison-Wesley, 1992).
2. Fitzgerald, Kingsley, and Umans, *Electric Machinery*, 5th ed. (McGraw Hill, 1990).

PROBLEMS

5.1 In Example 5.1, calculate the field intensity within the core: (a) very close to the inside diameter and (b) very close to the outside diameter. (c) Compare the results with the field intensity along the mean path.

5.2 In Example 5.1, calculate the reluctance in the path of flux lines if $\mu_r = 2000$.

5.3 Consider the core of dimensions given in Example 5.1. The coil requires an inductance of 25 μH. The maximum current is 3 A and the maximum flux density is not to exceed 1.3 T. Calculate the number of turns N and the relative permeability μ_r of the magnetic material that should be used.

5.4 In Problem 5.3, assume the permeability of the magnetic material to be infinite. To satisfy the conditions of maximum flux density and the desired inductance, a small air gap is introduced. Calculate the length of this air gap (neglecting the effect of flux fringing) and the number of turns N.

5.5 In Example 5.4, calculate the maximum current beyond which the flux density in the core will exceed 0.3 T.

5.6 The rectangular toroid of Figure 5.7 in Example 5.4 consists of a material whose relative permeability can be assumed to be infinite. The other parameters are as given in Example 5.4. An air gap of 0.05 mm length is introduced. Calculate (a) the coil inductance L_m assuming that the core is unsaturated and (b) the maximum current beyond which the flux density in the core will exceed 0.3 T.

5.7 In Problem 5.6, calculate the energy stored in the core and in the air gap at the flux density of 0.3 T.

5.8 In the structure of Figure 5.11a, $L_m = 200$ mH, $L_l = 1$ mH, and $N = 100$ turns. Ignore the coil resistance. A steady state voltage is applied, where $\overline{V} = \sqrt{2} \times 120 \angle 0V$ at a frequency of 60 Hz. Calculate the current $\overline{I}$ and i(t).

5.9 A transformer is designed to step down the applied voltage of 120 V (rms) to 24 V (rms) at 60 Hz. Calculate the maximum rms voltage that can be applied to the high-side of this transformer without exceeding the rated flux density in the core if this transformer is used in a power system with a frequency of 50 Hz.

5.10 Assume the transformer in Figure 5.15a to be ideal. Winding 1 is applied a sinusoidal voltage in steady state with $\overline{V}_1 = \sqrt{2} \times 120 \ V \angle 0°$ at a frequency $f = 60$ Hz. $N_1/N_2 = 3$. The load on winding 2 is a series combination of R and L with $Z_L = (5 + j3)\Omega$. Calculate the current drawn from the voltage source.

5.11 Consider the transformer shown in Figure 5.15a, neglecting the winding resistances, leakage inductances, and the core loss. $N_1/N_2 = 3$. For a voltage of 120 V (rms) at a frequency of 60 Hz applied to winding 1, the magnetizing current is 1.0 A (rms). If a load of 1.1 Ω at a power factor of 0.866 (lagging) is connected to the secondary winding, calculate $\overline{I}_1$.

5.12 In Problem 5.11, the core of the transformer now consists of a material with a μ_r that is one-half of that in Problem 5.11. Under the operating conditions listed in Problem 5.11, what are the core flux density and the magnetizing current? Compare these values to those in Problem 5.11. Calculate $\overline{I}_1$.

5.13 A 2400 / 240 V, 60-Hz transformer has the following parameters in the equivalent circuit of Figure 5.16: the high-side leakage impedance is $(1.2 + j2.0)$ Ω, the low-side leakage impedance is $(0.012 + j0.02)$ Ω, and X_m at the high-side is 1800 Ω. Neglect R_{he}. Calculate the input voltage if the output voltage is 240 V (rms) and supplying a load of 1.5 Ω at a power factor of 0.9 (lagging).

5.14 Calculate the equivalent-circuit parameters of a transformer, if the following open-circuit and short-circuit test data is given for a 60-Hz, 50-kVA, 2400:240 V distribution transformer:

open-circuit test with high-side open: $V_{oc} = 240$ V, $I_{oc} = 5.0$ A, $P_{oc} = 400$ W,
short-circuit test with low-side shorted: $V_{sc} = 90$ V, $I_{sc} = 20$ A, $P_{sc} = 700$ W

6

BASIC PRINCIPLES OF ELECTROMECHANICAL ENERGY CONVERSION

6.1 INTRODUCTION

Electric machines, as motors, convert electrical power input into mechanical output, as shown in Figure 6.1. These machines may be operated solely as generators, but they also enter the generating mode when slowing down (during regenerative braking) where the power flow is reversed. In this chapter, we will briefly look at the basic structure of electric machines and the fundamental principles of the electromagnetic interactions that govern their operation. We will limit our discussion to rotating machines, although the same principles apply to linear machines.

6.2 BASIC STRUCTURE

We often describe electric machines by viewing a cross-section, as if the machine were "cut" by a plane perpendicular to the shaft axis and viewed from one side, as shown in Figure 6.2a. Because of symmetry, this cross-section can be taken anywhere along the shaft axis. The simplified cross-section in Figure 6.2b shows that all machines have a stationary part, called the stator, and a rotating part, called the rotor, separated by an air gap, thereby allowing the rotor to rotate freely on a shaft, supported by bearings. The stator is firmly affixed to a foundation to prevent it from turning.

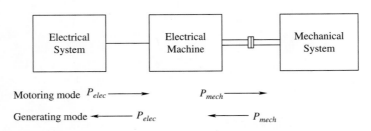

FIGURE 6.1 Electric machine as an energy converter.

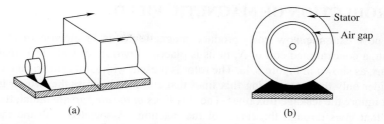

FIGURE 6.2   Motor construction (a) "cut" perpendicular to the shaft-axis; (b) cross-section seen from one side.

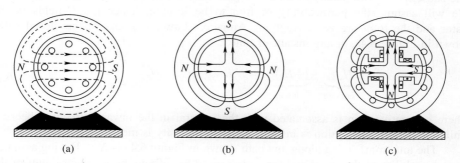

FIGURE 6.3   Structure of machines.

In order to require small ampere-turns to create flux lines shown crossing the air gap in Figure 6.3a, both the rotor and the stator are made up of high-permeability ferromagnetic materials, and the length of the air gap is kept as small as possible. In machines with ratings under 10 kW, a typical length of the air gap is about 1 mm, which is shown highly exaggerated for ease of drawing.

The stator-produced flux distribution in Figure 6.3a is shown for a 2-pole machine where the field distribution corresponds to a combination of a single north pole and a single south pole. Often there are more than 2 poles, for example 4 or 6. The flux-distribution in a 4-pole machine is represented in Figure 6.3b. Due to complete symmetry around the periphery of the airgap, it is sufficient to consider only one pole pair consisting of adjacent north and south poles. Other pole pairs have identical conditions of magnetic fields and currents.

If the rotor and the stator are perfectly round, the air gap is uniform, and the magnetic reluctance in the path of flux lines crossing the air gap is uniform. Machines with such structures are called non-salient pole machines. Sometimes, the machines are purposely designed to have saliency so that the magnetic reluctance is unequal along various paths, as shown in Figure 6.3c. Such saliency results in what is called the reluctance torque, which may be the primary or a significant means of producing the torque.

We should note that to reduce eddy-current losses, the stator and the rotor often consist of laminations of silicon steel, which are insulated from each other by a layer of thin varnish. These laminations are stacked together, perpendicular to the shaft axis. Conductors that run parallel to the shaft axis may be placed in slots cut into these laminations to place. Readers are urged to purchase a used dc motor and an induction motor and then take them apart to look at their construction.

6.3   PRODUCTION OF MAGNETIC FIELD

We will now examine how coils produce magnetic fields in electric machines. For illustration, a concentrated coil of N_s turns is placed in two stator slots 180° (called full-pitch) apart, as shown in Figure 6.4a. The rotor is present without its electrical circuit. We will consider only the magnetizing flux lines that completely cross the two air gaps, and at present ignore the leakage flux lines. The flux lines *in the air gap* are radial, that is, in a direction that goes through the center of the machine. Associated with the radial flux lines, the field intensity in the air gap is also in a radial direction; it is assumed to be positive $(+H_s)$ if it is away from the center of the machine, otherwise negative $(-H_s)$. The subscript "s" (for stator) refers to the field intensity *in the air gap* due to the stator. We will assume the permeability of iron to be infinite, hence the H-fields in the stator and the rotor are zero. Applying Ampere's Law along any of the closed paths shown in Figure 6.4a, at any instant of time t,

$$\underbrace{H_s\,\ell_g}_{\text{outward}} - \underbrace{(-H_s)\ell_g}_{\text{inward}} = N_s i_s \qquad \text{or} \qquad H_s = \frac{N_s i_s}{2\ell_g} \qquad (6.1)$$

where a negative sign is associated with the integral in the inward direction, because while the path of integration is inward, the field intensity is measured outward.

The total mmf acting along any path shown in Figure 6.4a is $N_s i_s$. Having assumed the permeability of the stator and the rotor iron to be infinite, by symmetry, half of the

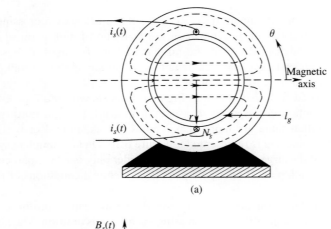

(a)

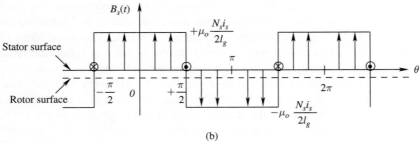

(b)

FIGURE 6.4   Production of magnetic field.

total ampere-turns $((N_s i_s)/2)$ are "consumed" or "acting" in making the flux lines cross each air gap length. Hence, the mmf F_s acting on each air gap is

$$F_s = \frac{N_s i_s}{2} \tag{6.2}$$

Substituting for $\frac{N_s i_s}{2}$ from Equation 6.2 into Equation 6.1,

$$F_s = H_s \, \ell_g \tag{6.3}$$

Associated with H_s in the air gap is the flux density B_s, which using Equation 6.1 can be written as

$$B_s = \mu_o H_s = \mu_o \frac{N_s i_s}{2\ell_g} \tag{6.4}$$

All field quantities (H_s, F_s, and B_s) directed away from the center of the machine are considered positive. Figure 6.4b shows the "developed" view, as if the circular cross-section in Figure 6.4a were flat. Note that the field distribution is a square wave. From Equations 6.1, 6.2, and 6.4, it is clear that all three stator-produced field quantities (H_s, F_s, and B_s) are proportional to the instantaneous value of the stator current $i_s(t)$ and are related to each other by constants. Therefore, in Figure 6.4b, the square wave plot of B_s distribution at an instant of time also represents H_s and B_s distributions at that time, plotted on different scales.

In the structure of Figure 6.4a, the axis through $\theta = 0°$ is referred to as the magnetic axis of the coil or winding that is producing this field. The magnetic axis of a winding goes through the center of the machine in the direction of the flux lines produced by a positive value of the winding current and is perpendicular to the plane in which the winding is located.

Example 6.1
In Figure 6.4a, consider a concentrated coil with $N_s = 25$ turns, and air gap length $\ell_g = 1$ mm. The mean radius (at the middle of the air gap) is $r = 15$ cm, and the length of the rotor is $\ell = 35$ cm. At an instant of time t, the current $i_s = 20$ A. (a) Calculate the H_s, F_s, and B_s distributions in the air gap as a function of θ, and (b) calculate the total flux crossing the air gap.

Solution

(a) Using Equation 6.2, $F_s = \frac{N_s i_s}{2} = 250$ A $\cdot$ turns. From Equation 6.1, $H_s = \frac{N_s i_s}{2\ell_g} = 2.5 \times 10^5$ A/m. Finally, using Equation 6.4, $B_s = \mu_o H_s = 0.314$ T. Plots of the field distributions are similar to those shown in Figure 6.4b.

(b) The flux crossing the rotor is $\phi_s = \int B \cdot dA$, calculated over half of the curved cylindrical surface A. The flux density is uniform and the area A is one-half of the circumference times the rotor length: $A = \frac{1}{2}(2\pi r)\ell = 0.165$ m^2. Therefore, $\phi_s = B_s \cdot A = 0.0518$ Wb.

Note that the length of the air gap in electrical machines is extremely small, typically one to two mm. Therefore, we will use the radius r at the middle of the air gap to also represent the radius to the conductors located in the rotor and the stator slots.

6.4 BASIC PRINCIPLES OF OPERATION

There are two basic principles that govern electric machines' operation to convert between electric energy and mechanical work:

(1) A force is produced on a current-carrying conductor when it is subjected to an *externally-established* magnetic field.
(2) An emf is induced in a conductor moving in a magnetic field.

6.4.1 Electromagnetic Force

Consider the conductor of length ℓ shown in Figure 6.5a. The conductor is carrying a current i and is subjected to an *externally-established* magnetic field of a uniform flux-density B perpendicular to the conductor length. A force f_{em} is exerted on the conductor due to the electromagnetic interaction between the external magnetic field and the conductor current. The magnitude of this force is given as

$$\underbrace{f_{em}}_{[Nm]} = \underbrace{B}_{[T]} \quad \underbrace{i}_{[A]} \quad \underbrace{\ell}_{[m]} \tag{6.5}$$

As shown in Figure 6.5a, the direction of the force is perpendicular to the directions of both i and B. To obtain the direction of this force, we will superimpose the flux lines produced by the conductor current, which are shown in Figure 6.5b. The flux lines add up on the right side of the conductor and subtract on the left side, as shown in Figure 6.5c. Therefore, the force f_{em} acts *from the higher concentration of flux lines to the lower concentration*, that is, from right to left in this case.

Example 6.2
In Figure 6.6a, the conductor is carrying a current into the paper plane in the presence of an external, uniform field. Determine the direction of the electromagnetic force.

Solution The flux lines are clockwise and add up on the upper-right side, hence the resulting force shown in Figure 6.6b.

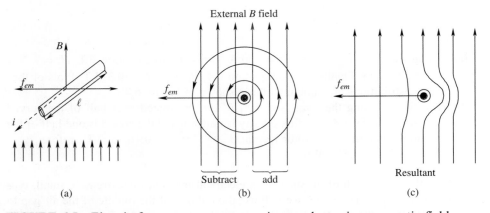

FIGURE 6.5 Electric force on a current-carrying conductor in a magnetic field.

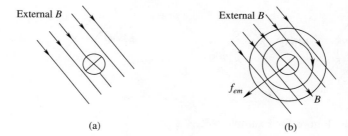

FIGURE 6.6 Figure for Example 6.2.

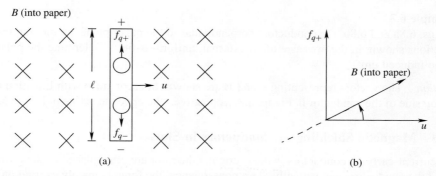

FIGURE 6.7 Conductor moving in a magnetic field.

6.4.2 Induced EMF

In Figure 6.7a, a conductor of length ℓ is moving to the right at a speed u. The B-field is uniform and is perpendicularly directed into the paper plane. The magnitude of the induced emf at any instant of time is then given by

$$\underbrace{e}_{[V]} = \underbrace{B}_{[T]} \underbrace{l}_{[m]} \underbrace{u}_{[m/s]} \tag{6.6}$$

The polarity of the induced emf can be established as follows: due to the conductor motion, the force on a charge q (positive, or negative in the case of an electron) within the conductor can be written as

$$f_q = q\,(\mathbf{u} \times \mathbf{B}) \tag{6.7}$$

where the speed and the flux density are shown by bold letters to imply that these are vectors and their cross product determines the force.

Since $\mathbf{u}$ and $\mathbf{B}$ are orthogonal to each other, as shown in Figure 6.7b, the force on a positive charge is upward. Similarly, the force on an electron will be downwards. Thus, the upper end will have a positive potential with respect to the lower end. This induced emf across the conductor is independent of the current that would flow if a closed path were to be available (as would normally be the case). With the current flowing, the

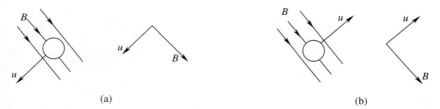

(a)   (b)

FIGURE 6.8   Figure for Example 6.3.

voltage across the conductor will be the induced-emf $e(t)$ in Equation 6.6 minus the voltage drops across the conductor resistance and inductance.

Example 6.3

In Figs. 6.8a and 6.8b, the conductors perpendicular to the paper plane are moving in the directions shown, in the presence of an external, uniform B-field. Determine the polarity of the induced emf.

Solution   The vectors representing **u** and **B** are shown. In accordance with Equation 6.7, the top side of the conductor in Figure 6.8a is positive. The opposite is true in Figure 6.8b.

6.4.3   Magnetic Shielding of Conductors in Slots

The current-carrying conductors in the stator and the rotor are often placed in slots, which shield the conductors magnetically. As a consequence, the force is mainly exerted on the iron around the conductor. It can be shown, although we will not prove it here, that this force has the same magnitude and direction as it would in the absence of the magnetic shielding by the slot. Since our aim in this course is not to design but rather to utilize electric machines, we will completely ignore the effect of the magnetic shielding of conductors by slots in our subsequent discussions. The same argument applies to the calculation of induced emf and its direction using Equations 6.6 and 6.7.

6.5   APPLICATION OF THE BASIC PRINCIPLES

Consider the structure of Figure 6.9a, where we will assume that the stator has established a uniform field B_s in the radial direction through the air gap. An N_r-turn coil is located on the rotor at a radius r. We will consider the force and the torque acting on the rotor in the counter-clockwise direction to be positive.

A current i_r is passed through the rotor coil, which is subjected to a stator-established field B_s in Figure 6.9a. The coil inductance is assumed to be negligible. The current magnitude I is constant, but its direction (details are discussed in Chapter 7) is controlled such that it depends on the location δ of the coil, as plotted in Figure 6.9b. In accordance with Equation 6.5, the force on both sides of the coil results in an electro-magnetic torque on the rotor in a counter-clockwise direction, where

$$f_{em} = B_s \left(N_r I \right) \ell \tag{6.8}$$

Thus,

$$T_{em} = 2 f_{em}\, r = 2 B_s \left(N_r I \right) \ell r \tag{6.9}$$

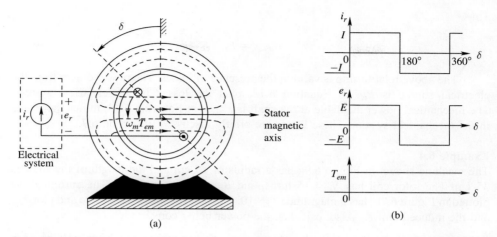

(a)

(b)

FIGURE 6.9 Motoring mode.

As the rotor turns, the current direction is changed every half-cycle, resulting in a torque that remains constant, as given by Equation 6.9. This torque will accelerate the mechanical load connected to the rotor shaft, resulting in a speed ω_m. Note that an equal but opposite torque is experienced by the stator. This is precisely the reason for affixing the stator to the foundation—to prevent the stator from turning.

Due to the conductors moving in the presence of the stator field, in accordance with Equation 6.6, the magnitude of the induced emf at any instant of time in each conductor of the coil is

$$E_{cond} = B_s \underbrace{\ell r \omega_m}_{u} \tag{6.10}$$

Thus, the magnitude of the induced emf in the rotor coil with $2N_r$ conductors is

$$E = 2N_r B_s \ell r \omega_m \tag{6.11}$$

The waveform of the emf e_r, with the polarity indicated in Figure 6.9a, is similar to that of i_r, as plotted in Figure 6.9b.

6.6 ENERGY CONVERSION

In this idealized system, which has no losses, we can show that the electrical input power P_{el} is converted into the mechanical output power P_{mech}. Using the waveforms of i_r, e_r, and T_{em} in Figure 6.9b at any instant of time,

$$P_{el} = e_r i_r = (2N_r B_s \ell r \omega_m) I \tag{6.12}$$

and,

$$P_{mech} = T_{em}\, \omega_m = (2B_s N_r I \ell r)\omega_m \tag{6.13}$$

Thus,

$$P_{mech} = P_{el} \tag{6.14}$$

The above relationship is valid in the presence of losses. The power drawn from the electrical source is P_{el}, in Equation 6.12, plus the losses in the electrical system. The mechanical power available at the shaft is P_{mech}, in Equation 6.13, minus the losses in the mechanical system. These losses are briefly discussed in section 6.7.

Example 6.4

The machine shown in Figure 6.9a has a radius of 15 cm and the length of the rotor is 35 cm. The rotor coil has $N_r = 15$ turns, and $B_s = 1.3\ T$ (uniform). The current i_r, as plotted in Figure 6.9b, has a magnitude $I = 10\ A$. $\omega_m = 100\ \text{rad/s}$. Calculate and plot T_{em} and the induced emf e_r. Also, calculate the power being converted.

Solution Using Equation 6.9, the electromagnetic torque on the rotor will be in a counter-clockwise direction and of a magnitude

$$T_{em} = 2B_s(N_r I)\ell r = 2 \times 1.3 \times 15 \times 10 \times 0.35 \times 0.15 = 20.5\ \text{Nm}$$

The electromagnetic torque will have the waveform shown in Figure 6.9b. At a speed of $\omega_m = 100\ \text{rad/s}$, the electrical power absorbed for conversion into mechanical power is

$$P = \omega_m T_{em} = 100 \times 20.5 \simeq 2\ \text{kW}$$

6.6.1 Regenerative Braking

At a speed ω_m, the rotor inertia, including that of the connected mechanical load, has stored kinetic energy. This energy can be recovered and fed back into the electrical system shown in Figure 6.10a.

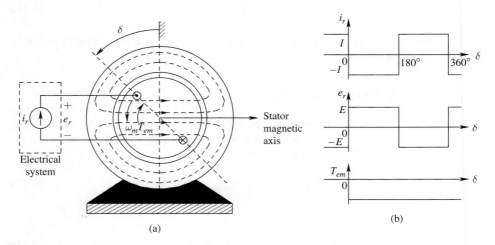

FIGURE 6.10   Regenerative braking mode.

In so doing, the current is so controlled as to have the waveform plotted in Figure 6.10b, as a function of the angle δ. Notice that the waveform of the induced voltage remains unchanged. In this regenerative case, due to the reversal of the current direction (compared to that in the motoring mode), the torque T_{em} is in the clockwise direction (opposing the rotation), and shown to be negative in Figure 6.10b. Now the input power P_{mech} from the mechanical side equals the output power P_{el} into the electrical system. This direction of power flow represents the generator mode of operation.

6.7 POWER LOSSES AND ENERGY EFFICIENCY

As indicated in Figure 6.11, any electric drive has inherent power losses, which are converted into heat. These losses, which are complex functions of the speed and the torque of the machine, are discussed in Chapter 15. If the output power of the drive is P_o, then the input power to the motor in Figure 6.11 is

$$P_{in,motor} = P_o + P_{loss,motor} \qquad (6.15)$$

At any operating condition, let $P_{loss, motor}$ equal all of the losses in the motor. Then the energy efficiency of the motor is

$$\eta_{motor} = \frac{P_o}{P_{in,motor}} = \frac{P_o}{P_o + P_{loss,motor}} \qquad (6.16)$$

In the power-processing unit of an electric drive, power losses occur due to current conduction and switching within the power semiconductor devices. Similar to Equation 6.16, we can define the energy efficiency of the PPU as η_{PPU}. Therefore, the overall efficiency of the drive is such that

$$\eta_{drive} = \eta_{motor} \times \eta_{PPU} \qquad (6.17)$$

Energy efficiencies of drives depend on many factors, which we will discuss in Chapter 15. The energy efficiency of small-to-medium-sized electric motors ranges from 85 to 93 percent, while that of power-processing units ranges from 93 to 97 percent. Thus, from Equation 6.17, the overall energy efficiency of drives is in the approximate range of 80 to 90 percent.

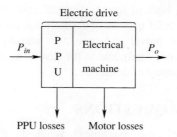

FIGURE 6.11 Power losses and energy efficiency.

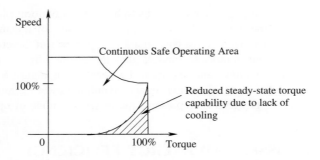

FIGURE 6.12 Safe operating area for electric machines.

6.8 MACHINE RATINGS

Machine ratings specify the limits of speed and torque at which a machine can operate. Usually they are specified for the continuous duty operation. These limits may be higher for intermittent duty and for dynamic operation during brief periods of accelerations and decelerations. Power loss in a machine raises its temperature above the temperature of its surroundings, which is often referred to as the ambient temperature. The ambient temperature is usually taken to be $40°C$. Machines are classified based on the temperature rise that they can tolerate. The temperature should not exceed the limit specified by the machine class. As a rule of thumb, operation at $10°C$ above the limit reduces the motor life expectancy by 50 percent.

The name-plate on the machine usually specifies the continuous-duty, full-load operating point in terms of the full-load torque, called the rated torque, and the full-load speed, called the rated speed. The product of these two values specifies the full-load power, or the rated power:

$$P_{rated} = \omega_{rated}\, T_{rated} \qquad (6.18)$$

The maximum speed of a motor is limited due to structural reasons such as the capability of the bearings and the rotor to withstand high speeds. The maximum torque that a motor can deliver is limited by the temperature rise within the motor. In all machines, higher torque output results in larger power losses. The temperature rise depends on power losses as well as cooling. In self-cooled machines, the cooling is not as effective at lower speeds; this reduces the machine torque capability at lower speeds. The torque-speed capability of electrical machines can be specified in terms of a safe operating area (SOA), as shown in Figure 6.12. The torque capability declines at lower speeds due to insufficient cooling. An expanded area, both in terms of speed and torque, is usually possible for intermittent duty and during brief periods of acceleration and deceleration.

In addition to the rated power and speed, the name-plate also specifies the rated voltage, the rated current (at full-load), and in the case of ac machines, the power factor at full load and the rated frequency of operation.

SUMMARY/REVIEW QUESTIONS

1. What is the role of electric machines? What do the motoring-mode and the generating-mode of operations mean?
2. What are the definitions of stator and rotor?

3. Why do we use high-permeability ferromagnetic materials for stators and rotors in electric machines? Why are these constructed by stacking laminations together, rather than as a solid structure?
4. What is the approximate air gap length in machines with less than 10 kW ratings?
5. What are multi-pole machines? Why can such machines be analyzed by considering only one pair of poles?
6. Assuming the permeability of iron to be infinite, where is the mmf produced by machine coils "consumed"? What law is used to calculate the field quantities, such as flux density, for a given current through a coil? Why is it important to have a small air gap length?
7. What are the two basic principles of operation for electric machines?
8. What is the expression for force acting on a current-carrying conductor in an externally established *B*-field? What is its direction?
9. What is slot shielding, and why can we choose to ignore it?
10. How do we express the induced emf in a conductor "cutting" an externally established *B*-field? How do we determine the polarity of the induced emf?
11. How do electrical machines convert energy from one form to another?
12. What are various loss mechanisms in electric machines?
13. How is electrical efficiency defined, and what are typical values of efficiencies for the machines, the power-processing units, and the overall drives?
14. What is the end-result of power losses in electric machines?
15. What is meant by the various ratings on the name-plates of machines?

REFERENCES

1. A. E. Fitzgerald, C. Kingsley and S. Umans, *Electric machinery*, 5th edition (McGraw-Hill, Inc., 1990).
2. G. R. Slemon, *Electric Machines and Drives* (Addison-Wesley, 1992).

PROBLEMS

6.1   Assume the field distribution produced by the stator in the machine shown in Figure P6.1 to be radially uniform. The magnitude of the air gap flux density is B_s, the rotor length is ℓ, and the rotational speed of the motor is ω_m.

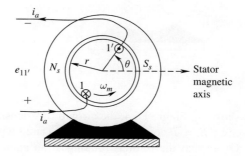

FIGURE P6.1

(a) Plot the emf $e_{11'}$ induced in the coil as a function of θ for two values of $i_a : 0$ A and 10 A.

(b) In the position shown, the current i_a in the coil $11'$ equals I_o. Calculate the torque acting on the coil in this position for two values of instantaneous speed $\omega_m : 0\,\text{rad}/\text{s}$ and $100\,\text{rad}/\text{s}$.

6.2   Figure P6.2 shows a primitive machine with a rotor producing a uniform magnetic field such that the air-gap flux density in the radial direction is of the magnitude B_r. Plot the induced emf $e_{11'}$ as a function of θ. The length of the rotor is ℓ and the radius at the air gap is r.

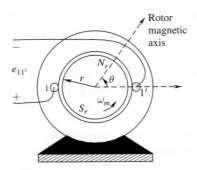

FIGURE P6.2

6.3   In the primitive machine shown in Figure P6.3, the air-gap flux density B_s has a sinusoidal distribution given by $B_S = \hat{B} \cos \theta$. The rotor length is ℓ. (a) Given that the rotor is rotated at a speed of ω_m, plot as a function of θ the emf $e_{11'}$ induced and the torque T_{em} acting on the coil if $i_a = I$. (b) In the position shown, the current i_a in the coil equals I. Calculate P_{el}, the electrical power input to the machine, and P_{mech}, the mechanical power output of the machine, if $\omega_m = 60$ rad/s.

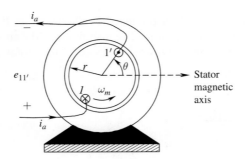

FIGURE P6.3

6.4   In the machine shown in Figure P6.4, the air-gap flux density B_r has a sinusoidal distribution given by $B_r = \hat{B} \cos \theta$, where α is measured with respect to the rotor magnetic axis. Given that the rotor is rotating at an angular speed ω_m and the rotor length is ℓ, plot the emf $e_{11'}$ induced in the coil as a function of θ.

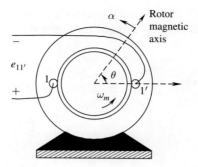

FIGURE P6.4

6.5   In the machine shown in Figure P6.5, the air gap flux density B_s is constant and equal to B_{max} in front of the pole faces, and is zero elsewhere. The direction of the *B*-field is from left (north pole) to right (south pole). The rotor is rotating at an angular speed of ω_m and the length of the rotor is ℓ. Plot the induced emf $e_{11'}$ as a function of θ. What should be the waveform of i_a that produces an optimum electromagnetic torque T_{em}?

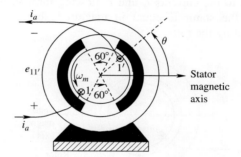

FIGURE P6.5

6.6   As shown in Figure P6.6, a rod in a uniform magnetic field is free to slide on two rails. The resistances of the rod and the rails are negligible. Electrical continuity between the rails and the rod is assumed, so a current can flow through the rod. A damping force, F_d, tending to slow down the rod, is proportional to the square of the rod's speed as follows: $F_d = k_f u^2$ where $k_f = 1500$. Assume that the inductance in this circuit can be ignored. Find the steady state speed u of the rod, assuming the system extends endlessly to the right.

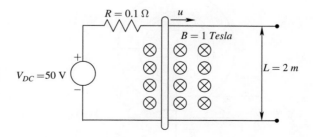

FIGURE P6.6

6.7   Consider Figure P6.7. Plot the mmf distribution in the air gap as a function of θ for $i_a = I$. Assume each coil has a single turn.

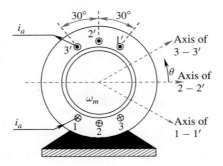

FIGURE P6.7

6.8   In Figure P6.8, the stator coil has N_s turns and the rotor coil has N_r turns. Each coil produces in the air gap a uniform, radial flux density B_s and B_r, respectively. In the position shown, calculate the torque experienced by both the rotor coil and the stator coil, due to the currents i_s and i_r flowing through these coils. Show that the torque on the stator is equal in magnitude but opposite in direction to that experienced by the rotor.

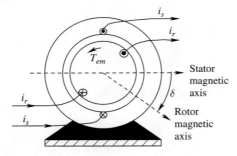

FIGURE P6.8

6.9   Figure P6.9 shows the cross-section, seen from the front, of squirrel-cage induction machines, which are discussed in Chapter 11. In such machines, the magnetizing-flux density distribution produced by the stator windings is

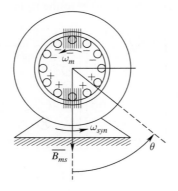

FIGURE P6.9

represented by a vector $\vec{B}_{ms}$ that implies that at this instant, for example, the radially-oriented flux density in the air gap is in a vertically-downward direction where the peak occurs, and it is co-sinusoidally distributed elsewhere. This means that at this instant, the flux-density in the air gap is zero at $\theta = 90°$ (and at $\theta = 270°$). At $\theta = 180°$, the flux-density again peaks but has a negative value. This flux-density distribution is rotating at a speed ω_{syn}. The rotor consists of bars, as shown, along its periphery. This rotor is rotating at a speed of ω_m. The voltages induced in the front-ends of the rotor-bars have the polarities as shown in Figure P6.9, with respect to their back-ends. Determine if ω_m is greater or less than ω_{syn}.

7

DC-MOTOR DRIVES AND ELECTRONICALLY-COMMUTATED MOTOR (ECM) DRIVES

7.1 INTRODUCTION

Historically, dc-motor drives have been the most popular drives for speed and position control applications. They owe this popularity to their low cost and ease of control. Their demise has been prematurely predicted for many years. However, they are losing their market share to ac drives due to wear in their commutator and brushes, which require periodic maintenance. Another factor in the decline of the market share of dc drives is cost. Figure 7.1 shows the cost distribution within dc drives in comparison to ac drives at present and in the future. In inflation-adjusted dollars, the costs of ac and dc motors are expected to remain nearly constant. For power processing and control, ac drives require more complex electronics (PPUs), making them at present more expensive than in dc-motor drives. However, the cost of drive electronics (PPUs) continues to decrease. Therefore, ac drives are gaining market share over dc drives.

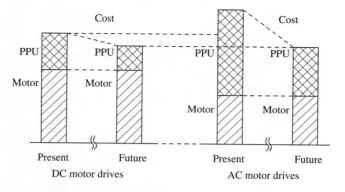

FIGURE 7.1 Cost distribution with dc and ac drives.

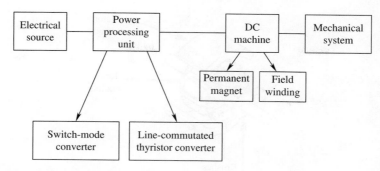

FIGURE 7.2   Classification of DC drives.

There are two important reasons to learn about dc drives. First, a number of such drives are currently in use, and this number keeps increasing. Second, the control of ac-motor drives emulates the operating principles of dc-motor drives. Therefore, knowledge of dc-motor drives forms the first step in learning how to control ac drives.

In a dc-motor drive, dc voltage and current are supplied by a power-processing unit to the dc motor, as shown in the block diagram of Figure 7.2. There are two designs of dc machines: stators consisting of either permanent magnets or a field winding. The power-processing units can also be classified into two categories: switch-mode power converters that operate at a high switching frequency, as discussed in Chapter 4, or line-commutated, thyristor converters, which are discussed later in Chapter 16. In this chapter, our focus will be on small servo-drives, which usually consist of permanent-magnet motors supplied by switch-mode power electronic converters.

At the end of this chapter, a brief discussion of Electronically-Commutated Motors (ECM) is included as a way of reinforcing the concept of current commutation, as well as a way of introducing an important class of motor drives that do not have the problem of wear in commutator and brushes. Therefore, in the trade literature, ECM are also referred as brush-less dc drives.

7.2   THE STRUCTURE OF DC MACHINES

Figure 7.3 shows a cut-away view of a dc motor. It shows a permanent-magnet stator, a rotor which carries the armature winding, a commutator, and the brushes.

In dc machines, the stator establishes a uniform flux ϕ_f in the air gap in the radial direction (the subscript "f" is for field). If permanent magnets like those shown in the cross-section of Figure 7.4a are used, the air gap flux density established by the stator remains constant (it cannot be changed). A field winding whose current can be varied can be used to achieve an additional degree of control over the air gap flux density, as shown in Figure 7.4b.

Figures 7.3 and 7.5 show that the rotor slots contain a winding, called the armature winding, which handles electrical power for conversion to (or from) mechanical power at the rotor shaft. In addition, there is a commutator affixed to the rotor. On its outer surface, the commutator contains copper segments that are electrically insulated from each other by means of mica or plastic. The coils of the armature winding are connected

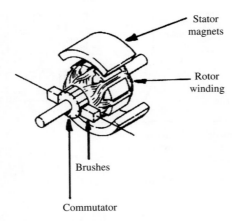

FIGURE 7.3   Exploded view of a dc motor [5].

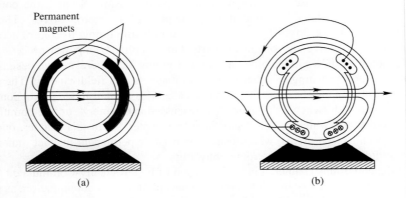

FIGURE 7.4   Cross-sectional view of magnetic field produced by stator.

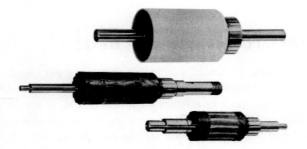

FIGURE 7.5   DC motor armatures [5].

to these commutator segments so that a stationary dc source can supply voltage and current to the rotating commutator by means of stationary carbon brushes that rest on top of the commutator. The wear due to the mechanical contact between the commutator and the brushes requires periodic maintenance, which is the main drawback of dc machines.

7.3 OPERATING PRINCIPLES OF DC MACHINES

The basic principle that governs the production of a steady electromagnetic torque has already been introduced in Chapter 6. A rotor coil in a uniform radial field established by the stator was supplied with a current, which reversed direction every half-cycle of rotation. The induced emf in the coil also alternated every half-cycle.

In practice, this reversal of current can be realized in the dc machine (still primitive) shown in Figure 7.6a, using two commutator segments (s_1 and s_2) and two brushes (b_1 and b_2). Using the notations commonly adopted in the context of dc machines, the armature quantities are indicated by the subscript "a", and the density of the stator-established flux (*that crosses the two air gaps*) is called the field flux density B_f, whose distribution as function of θ in Figure 7.6b is plotted in Figure 7.6c. In the plot of Figure 7.6c, the uniform flux density B_f in the air gap is assumed to be positive under the south pole and negative under the north pole. There is also a small "neutral" zone where the flux density is small and is changing from one polarity to the other.

We will see how the commutator and the brushes in the primitive (non-practical) machine of Figure 7.6a convert a dc current i_a supplied by a stationary source into an alternating current in the armature coil. The cross-sectional view of this primitive machine, looking from the front, is represented in Figure 7.7. For the position of the coil at $\theta = 0°$ shown in Figure 7.7a, the coil current $i_{1-1'}$ is positive and a counter-clockwise force is produced on each conductor.

Figure 7.7b shows the cross-section when the rotor has turned counter-clockwise by $\theta = 90°$. The brushes are wider than the insulation between the commutator segments. Therefore, in this elementary machine, the current i_a flows through the commutator segments, and no current flows through the conductors. In this region, the coil undergoes "commutation" where its current direction reverses as the rotor turns further. Figure 7.7c shows the cross-section at the rotor position $\theta = 180°$. Compared to Figure 7.7a at

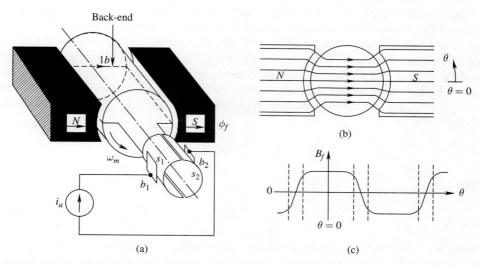

FIGURE 7.6 Flux density in the air gap.

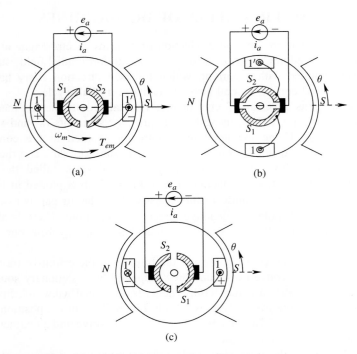

FIGURE 7.7   Torque production and commutator action.

$\theta = 0°$, the roles of conductors 1 and 1' are interchanged; hence at $\theta = 180°$, $i_{1-1'}$ is negative and the same counter-clockwise torque as at $\theta = 0°$ is produced.

The above discussion shows how the commutator and brushes convert a dc current at the armature terminals of the machine into a current that alternates every half-cycle through the armature coil. In the armature coil, the induced emf also alternates every half-cycle and is "rectified" at the armature terminals. The current and the induced emf in the coil are plotted in Figure 7.8a as a function of the rotor position θ. The torque on the rotor and the induced emf appearing at the brush terminals are plotted in Figures 7.8b and 7.8c where their average values are indicated by the dotted lines. Away from the "neutral zone," the torque and the induced emf expressions, in accordance with Chapter 6, are as follows:

$$T_{em} = (2B_f \ell r)i_a \qquad (7.1)$$

and

$$e_a = (2B_f \ell r)\omega_m \qquad (7.2)$$

where ℓ is effective conductor length and r is the radius. Notice the pronounced dip in the torque and the induced emf waveforms. These waveforms are improved by having a large number of distributed coils in the armature, as illustrated in the following example.

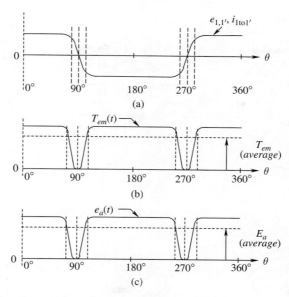

FIGURE 7.8   Waveforms for the motor in Figure 7.7.

Example 7.1

Consider the elementary dc machine shown in Figure 7.9 whose stator poles produce a uniform, radial flux density B_f in the air gap. The armature winding consists of 4 coils {1-1', 2-2', 3-3', and 4-4'} in 4 rotor slots. A dc current i_a is applied to the armature as shown. Assume the rotor speed to be ω_m (rad/s). Plot the induced emf across the brushes and the electromagnetic torque T_{em} as a function of the rotor position θ.

Solution   Figure 7.9 shows three rotor positions measured in a counter-clockwise (CCW) direction: $\theta = 0, 45°$ and $90°$. This figure shows how coils 1 and 3 go through current commutation. At $\theta = 0°$, the currents are from 1 to 1' and from 3' to 3. At $\theta = 45°$, the currents in these coils are zero. At $\theta = 90°$, the two currents have reversed direction. The total torque and the induced emf at the brush terminals are plotted in Figure 7.10.

If we compare the torque T_{em} and the emf e_a waveforms of the 4-coil winding in Figure 7.10 to those for the 1-coil winding in Figure 7.8, it is clear that pulsations in the torque and in the induced emf are reduced by increasing the number of coils and slots.

Practical dc machines consist of a large number of coils in their armature windings. Therefore, we can neglect the effect of the coils in the "neutral" zone undergoing current commutation, and the armature can be represented as shown in Figure 7.11.

The following conclusions regarding the commutator action can be drawn:

- The armature current i_a supplied through the brushes divides equally between two circuits connected in parallel. Each circuit consists of half of the total conductors, which are connected in series. All conductors under a pole have currents in the same direction. The respective forces produced on each conductor are in the same direction and add up to yield the total torque. The direction of the armature current i_a determines the direction of currents through the conductors. (The current direction is independent of the direction of rotation.) Therefore, the direction of the electromagnetic torque produced by the machine also depends on the direction of i_a.

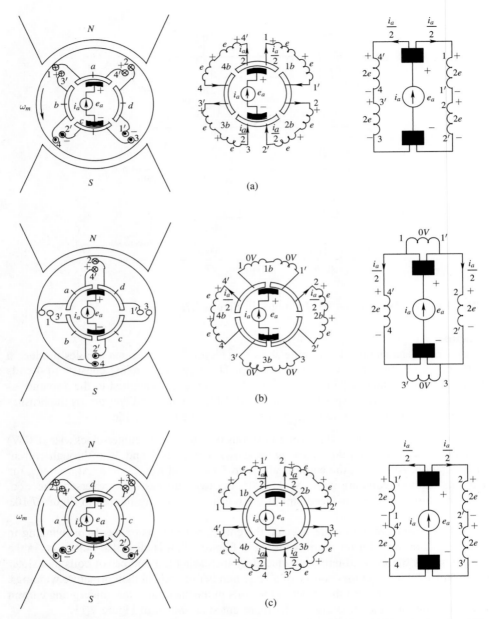

FIGURE 7.9 The dc machine in Example 7.1 (a) at $\theta = 0°$; (b) CCW rotation by 45°; (c) CCW rotation by 90°.

- The induced voltage in each of the two parallel armature circuits, and therefore across the brushes, is the sum of the voltages induced in all conductors connected in series. All conductors under a pole have induced emfs of the same polarity. The polarity of these induced emfs depends on the direction of rotation. (The emf polarity is independent of the current direction.)

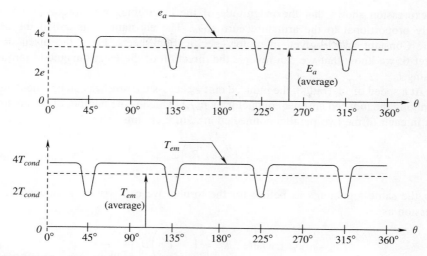

FIGURE 7.10   Torque and emf for Example 7.1.

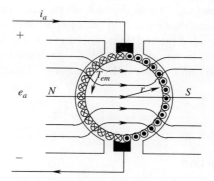

FIGURE 7.11   DC machine schematic representation.

We can now calculate the net torque produced and the emf induced. In the dc machine represented in Figure 7.11, let there be a total of n_a conductors, each of length l, placed in a uniform and radial flux density B_f. Then, the electromagnetic torque produced by a current $i_a/2$ can be calculated by multiplying the force per conductor by the number of conductors and the radius r:

$$T_{em} = \left(n_a \ell\, r\, B_f\right) \frac{i_a}{2} \qquad (7.3)$$

In a machine, the values of n_a, ℓ, and r are fixed. The flux density B_f also has a fixed value in a permanent-magnet machine. Therefore, we can write the torque expression as

$$T_{em} = k_T i_a \quad \text{where} \quad k_T = \left(\frac{n_a}{2}\ell\, r\right) B_f \left[\frac{\text{Nm}}{A}\right] \qquad (7.4)$$

This expression shows that the magnitude of the electromagnetic torque produced is linearly proportional to the armature current i_a. The constant k_T is called the Motor Torque Constant and is given in motor specification sheets. From the discussion in Chapter 6, we know that we can reverse the direction of the electromagnetic torque by reversing i_a.

At a speed of ω_m (rad/s), the induced emf e_a across the brushes can be calculated by multiplying the induced emf per conductor by $n_a/2$, which is the number of conductors in series in each of the two parallel connected armature circuits. Thus,

$$e_a = \left(\frac{n_a}{2} \ell r B_f \right) \omega_m \tag{7.5}$$

Using the same arguments as before for the torque, we can write the induced voltage expression as

$$e_a = k_E\, \omega_m \quad \text{where} \quad k_E = \left(\frac{n_a}{2} \ell r \right) B_f \left[\frac{V}{\text{rad/s}} \right] \tag{7.6}$$

This shows that the magnitude of the induced emf across the brushes is linearly proportional to the rotor speed ω_m. It also depends on the constant k_E, which is called the Motor Voltage Constant and is specified in motor specification sheets. The polarity of this induced emf is reversed if the rotational speed ω_m is reversed.

We should note that in any dc machine, the torque constant k_T and the voltage constant k_E are exactly the same, as shown by Equations 7.4 and 7.6, provided that we use the MKS units:

$$k_T = k_E = \left(\frac{n_a}{2} \ell r \right) B_f \tag{7.7}$$

7.3.1 Armature Reaction

Figure 7.12a shows the flux lines ϕ_f produced by the stator. The armature winding on the rotor, with i_a flowing through it, also produces flux lines, as shown in Figure 7.12b. These two sets of flux lines—ϕ_f and the armature flux ϕ_a—are at a right angle to each other. Assuming that the magnetic circuit does not saturate, we can superimpose the two sets of flux lines and show the combined flux lines in the air gap as in Figure 7.12c.

The fluxes ϕ_a and ϕ_f add in certain portions and subtract in the other portions. If the magnetic saturation is neglected as we have assumed, then due to the symmetry of

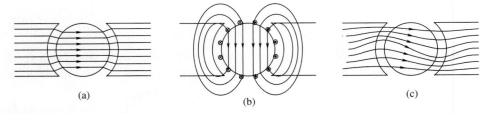

(a)   (b)   (c)

FIGURE 7.12   Effect of armature reaction.

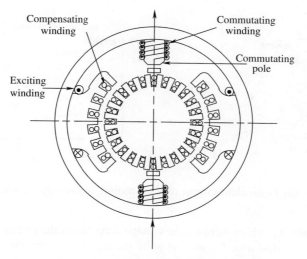

FIGURE 7.13   Measures to counter armature reaction.

the machine, the effect of an increased torque produced by conductors under higher flux density is canceled out by decreased torque produced by conductors under lower flux density. The same holds true for the induced emf e_a. Therefore, the calculations of the torque T_{em} and the induced emf e_a in the previous section remain valid.

If ϕ_a is so high that the net flux may saturate portions of the magnetic material in its path, then the superposition of the previous section is not valid. In that case, at high values of ϕ_a the net flux in the air gap near the saturated magnetic portions will be reduced compared to its value obtained by superposition. This will result in degradation in the torque produced by the given armature current. This effect is commonly called "saturation due to armature reaction." In our discussion we can neglect magnetic saturation and other ill-effects of armature reaction because in permanent-magnet machines the mmf produced by the armature winding results in a small ϕ_a. This is because there is a high magnetic reluctance in the path of ϕ_a. In field-wound dc machines, countermeasures can be taken: the mmf produced by the armature winding can be neutralized by passing the armature current in the opposite direction through a compensating winding placed on the pole faces of the stator, and through the commutating-pole windings, as shown in Figure 7.13.

7.4 DC-MACHINE EQUIVALENT CIRCUIT

It is often convenient to discuss a dc machine in terms of its equivalent circuit of Figure 7.14a, which shows conversion between electrical and mechanical power. In this figure, an armature current i_a is flowing. This current produces the electromagnetic torque $T_{em}(= k_T i_a)$ necessary to rotate the mechanical load at a speed of ω_m. Across the armature terminals, the rotation at the speed of ω_m induces a voltage, called the back-emf $e_a(= k_E \omega_m)$.

On the electrical side, the applied voltage v_a overcomes the back-emf e_a and causes the current i_a to flow. Recognizing that there is a voltage drop across both the armature

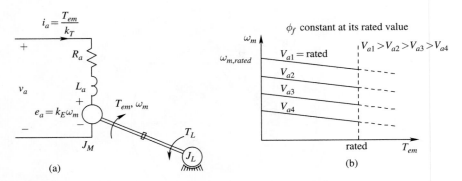

FIGURE 7.14   (a) Equivalent circuit of a dc motor; (b) steady state characteristics.

winding resistance R_a (which includes the voltage drop across the carbon brushes) and the armature winding inductance L_a, we can write the equation of the electrical side as

$$v_a = e_a + R_a\, i_a + L_a\frac{di_a}{dt} \tag{7.8}$$

On the mechanical side, the electromagnetic torque produced by the motor overcomes the mechanical-load torque T_L to produce acceleration:

$$\frac{d\omega_m}{dt} = \frac{1}{J_{eq}}(T_{em} - T_L) \tag{7.9}$$

where J_{eq} is the total effective value of the combined inertia of the dc machine and the mechanical load.

Note that the equations of the electric system and the mechanical system are coupled. The back-emf e_a in the electrical-system equation (Equation 7.8) depends on the mechanical speed ω_m. The torque T_{em} in the mechanical-system equation (Equation 7.9) depends on the electrical current i_a. The electrical power absorbed from the electrical source by the motor is converted into mechanical power and vice versa. In steady state, with a voltage V_a applied to the armature terminals, and a load-torque T_L supplied as well,

$$I_a = \frac{T_{em}(= T_L)}{k_T} \tag{7.10}$$

Also,

$$\omega_m = \frac{E_a}{k_E} = \frac{V_a - R_a I_a}{k_E} \tag{7.11}$$

The steady state torque-speed characteristics for various values of V_a are plotted in Figure 7.14b.

Example 7.2
A permanent-magnet dc motor has the following parameters: $R_a = 0.35\,\Omega$ and $k_E = k_T = 0.5$ in MKS units. For a torque of up to 8 Nm, plot its steady state torque-speed characteristics for the following values of V_a: 100 V, 75 V, and 50 V.

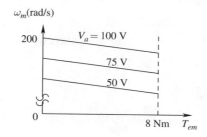

FIGURE 7.15 Example 7.3.

Solution Let's consider the case of $V_a = 100$ V. Ideally, at no-load (zero torque), from Equation 7.10, $I_a = 0$. Therefore, from Equation 7.11, the no-load speed is

$$\omega_m = \frac{V_a}{k_E} = \frac{100}{0.5} = 200 \text{ rad/s}$$

At a torque of 8 Nm, from Equation 7.10, $I_a = \frac{8\text{Nm}}{0.5} = 16A$. Again using Equation 7.11,

$$\omega_m = \frac{100 - 0.35 \times 16}{0.5} = 188.8 \text{ rad/s}$$

The torque-speed characteristic is a straight line, as shown in Figure 7.15. Similar characteristics can be drawn for the other values of V_a: 75 V and 50 V.

7.5 VARIOUS OPERATING MODES IN DC-MOTOR DRIVES

The major advantage of a dc-motor drive is the ease with which torque and speed can be controlled. A dc drive can easily be made to operate as a motor or as a generator in forward or reverse direction of rotation. In our prior discussions, the dc machine was operating as a motor in a counter-clockwise direction (which we will consider to be the forward direction). In this section, we will see how a dc machine can be operated as a generator during regenerative braking and how its speed can be reversed.

7.5.1 Regenerative Braking

Today, dc machines are seldom used as generators per se, but they operate in the generator mode in order to provide braking. For example, regenerative braking is used to slow the speed of a dc-motor-driven electric vehicle (most of which use PMAC drives, discussed in Chapter 10, but the principle of regeneration is the same) by converting kinetic energy associated with the inertia of the vehicle into electrical energy, which is fed into the batteries.

Initially, let's assume that a dc machine is operating in steady state as a motor and rotating in the forward direction as shown in Figure 7.16a.

A positive armature voltage v_a, which overcomes the back-emf e_a, is applied, and the current i_a flows to supply the load torque. The polarities of the induced emfs and the directions of the currents in the armature conductors are also shown.

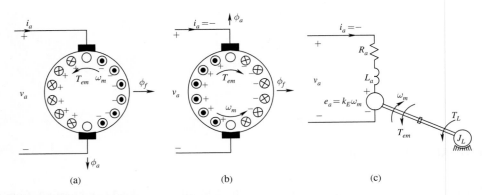

FIGURE 7.16   Regenerative braking.

One way of slowing down this dc-motor-driven electric vehicle is to apply mechanical brakes. However, a better option is to let the dc machine go into the generator mode by reversing the direction in which the electromagnetic torque T_{em} is produced. This is accomplished by reversing the direction of the armature current, which is shown to have a negative value in Figure 7.16b. Since the machine is still turning in the same direction (forward and counter-clockwise), the induced back-emf e_a remains positive. The armature-current direction can be reversed by decreasing the applied voltage v_a in comparison to the back-emf e_a (that is, $v_a < e_a$). The current reversal through the conductors causes the torque to reverse and oppose the rotation. Now, the power from the mechanical system (energy stored in the inertia) is converted and supplied to the electrical system. The equivalent circuit of Figure 7.16c in the generator mode shows the power being supplied to the electrical source (batteries, in the case of electrical vehicles).

Note that the torque $T_{em} (= k_T i_a)$ depends on the armature current i_a. Therefore, the torque will change as quickly as i_a is changed. DC motors for servo applications are designed with a low value of the armature inductance L_a; therefore, i_a and T_{em} can be controlled very quickly.

Example 7.3
Consider the dc motor of Example 7.2 whose moment-of-inertia $J_m = 0.02\ \text{kg} \cdot \text{m}^2$. Its armature inductance L_a can be neglected for slow changes. The motor is driving a load of inertia $J_L = 0.04\ \text{kg} \cdot \text{m}^2$. The steady state operating speed is 300 rad/s. Calculate and plot the $v_a(t)$ that is required to bring this motor to a halt as quickly as possible, without exceeding the armature current of 12 A.

Solution   In order to bring the system to a halt as quickly as possible, the maximum allowed current should be supplied, that is $i_a = -12\ A$. Therefore, $T_{em} = k_E i_a = -6$ Nm. The combined equivalent inertia is $J_{eq} = 0.06\ kg \cdot m^2$. From Equation 7.9 for the mechanical system,

$$\frac{d\omega_m}{dt} = \frac{1}{0.06}(-6.0) = -100 \text{ rad/s}.$$

Therefore, the speed will reduce to zero in 3 s in a linear fashion, as plotted in Figure 7.17. At time $t = 0^+$,

$$E_a = k_E \omega_m = 150 \text{ V}$$

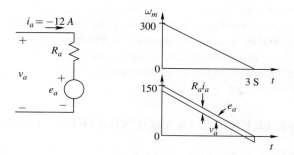

FIGURE 7.17 Example 7.4

and

$$V_a = E_a + R_aI_a = 150 + 0.35(-12) = 145.8 \text{ V}$$

Both e_a and v_a linearly decrease with time, as shown in Figure 7.17.

7.5.2 Operating in the Reverse Direction

Applying a reverse–polarity dc voltage to the armature terminals makes the armature current flow in the opposite direction. Therefore, the electromagnetic torque and the motor speed will also be reversed. Just as for the forward direction, regenerative braking is possible during rotation in the reverse direction.

7.5.3 Four-Quadrant Operation

As illustrated in Figure 7.18, a dc machine can be easily operated in all four quadrants of its torque-speed plane. For example, starting with motoring in the forward direction in the

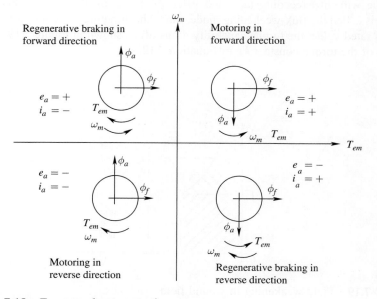

FIGURE 7.18 Four-quadrant operation.

upper-right section, it can be made to go into the other quadrants of operation by reversing the armature current and then reversing the applied armature voltage. In the upper-left quadrant, the drive is in the regenerative braking mode, while still running in the forward direction. In the lower-left quadrant, the drive is in the motoring mode in the reverse direction, while in the lower-right quadrant, it's in the regenerative braking mode in the reverse direction.

7.6 FLUX WEAKENING IN WOUND-FIELD MACHINES

In dc machines with a wound field, the field flux ϕ_f and the flux density B_f can be controlled by adjusting the field-winding current I_f. This changes the machine torque-constant and the voltage-constant given by Equations 7.4 and 7.6, both of which can be written explicitly in terms of B_f as

$$k_T = k_t B_f \tag{7.12}$$

and

$$k_E = k_e B_f \tag{7.13}$$

where the constants k_t and k_e are also equal to each other.

Below the rated speed, we will always keep the field flux at its rated value so that the torque-constant k_T is at its maximum value, which minimizes the current required to produce the desired torque, thus minimizing $i^2 R$ losses. At the rated field flux, the induced back-emf reaches its rated value at the rated speed. This is shown in Figure 7.19 as the constant-torque region.

What if we wish to operate the machine at speeds higher than the rated value? This would require a terminal voltage higher than the rated value. To work around this, we can reduce the field flux, which allows the motor to be operated at speeds higher than the rated value without exceeding the rated value of the terminal voltage. This mode of operation is called the flux-weakening mode. Since the armature current is not allowed to exceed its rated value, the torque capability drops off as shown in Figure 7.19 due to the reduction of the torque constant k_T in Equation 7.12.

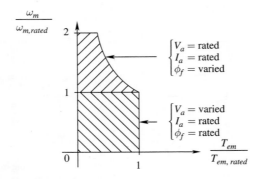

FIGURE 7.19   Field weakening in wound field machines.

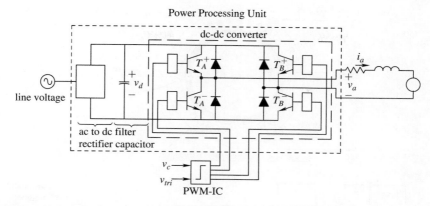

FIGURE 7.20   Switch-mode converter based PPU for DC motor drives.

7.7   POWER-PROCESSING UNITS IN DC DRIVES

In dc drives, the power-processing unit provides dc voltage and current to the armature of the dc machine. In general, this unit should be very energy efficient and should have a low cost. Depending on its application, the dc drive may be required to respond quickly and may also be operated in all 4-quadrants of Figure 7.18. Therefore, both v_a and i_a must be adjustable and reversible and independent of each other.

In most cases, the power-processing unit shown in Figure 7.2 is an interface between the electric utility and the dc machine (notable exceptions are vehicles supplied by batteries). Therefore, the power processor must draw power from the utility without causing or being susceptible to power quality problems. Ideally, the power flow through the PPU should be reversible into the utility system. The PPU should provide voltage and current to the dc machine with waveforms as close to a dc as possible. Deviations from a pure dc in the current waveform result in additional losses within the dc machine.

Power-processing units that utilize switch-mode conversion have already been discussed in Chapter 4. Their block diagram is repeated in Figure 7.20. The "front-end" of such units is usually a diode-rectifier bridge (which will be discussed in detail in Chapter 16). It is possible to replace the diode-rectifier "front-end" with a switch-mode converter to make the power flow *into* the utility during regenerative braking. The design of the feedback controller for dc drives will be discussed in detail in Chapter 8.

7.8   ELECTRONICALLY-COMMUTATED MOTOR (ECM) DRIVES

Earlier in this chapter, we have seen that the role of the commutator and the brushes is to reverse the direction of current through a conductor based on its location. The current through a conductor is reversed as it moves from one pole to the other. In brush-type dc motors discussed previously, the field flux is created by permanent magnets (or a field winding) on the stator, while the power-handling armature winding is on the rotor.

In contrast, in Electronically-Commutated Motors (ECM), the commutation of current is provided electronically, based on the positional information obtained from a sensor. These are "inside-out" machines where the magnetic field is established by the

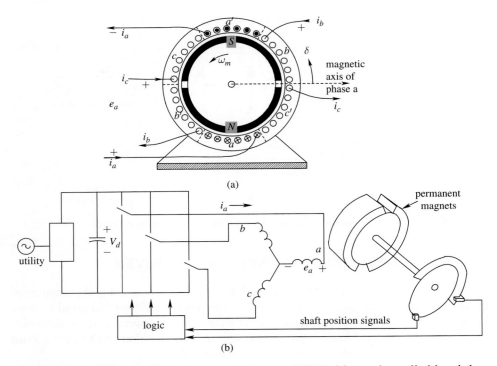

FIGURE 7.21 Electronically-commutated motor (ECM) drives; also called brush-less dc drives.

permanent magnets located on the rotor and the power handling winding is placed on the stator, as shown in Figure 7.21a. The block diagram of the drive, including the PPU and the position sensor, is shown in Figure 7.21b.

The stator in Figure 7.21a contains three-phase windings, which are displaced by 120 degrees. We will concentrate only on phase-*a*, because the roles of the other two phases are identical. The phase-*a* winding spans 60 degrees on each side, thus a total of 120 degrees, as shown in Figure 7.21a. It is connected in a wye-arrangement with the other phases, as shown in Figure 7.21b. It is distributed uniformly in slots, with a total of $2N_s$ conductors, where all of the conductors of the winding are in series. We will assume that the rotor produces a uniform flux density B_f distribution of flux lines crossing the air gap, rotating at a speed ω_m in a counter-clockwise direction. The flux-density distribution established by the rotor is rotating, but the conductors of the stator windings are stationary. The principle of the induced emf $e = B\ell u$ discussed in Chapter 6 is valid here as well. This is confirmed by the example below.

Example 7.4
Show that the principle $e = B\ell u$ applies to situations in which the conductors are stationary but the flux-density distribution is rotating.

Solution In Figure 7.21a, take a conductor from the top group and one from the bottom group at 180 degrees, forming a coil, as shown in Figure 7.22a. Figure 7.22b shows that the flux linkage of the coil is changing as a function of the rotor position δ (with $\delta = 0$ at the position shown in Figure 7.22a).

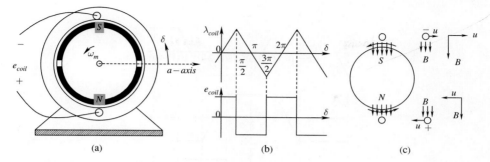

(a)   (b)   (c)

FIGURE 7.22   Example 7.5.

The peak flux linkage of the coil occurs at $\delta = \pi/2$ radians;

$$\hat{\lambda}_{coil} = (\pi r l)B_f \tag{7.14}$$

where ℓ is the rotor length and r is the radius. From Faraday's Law, the coil voltage equals the rate of change of the flux linkage. Therefore, recognizing that $\frac{d\delta}{dt} = \omega_m$,

$$e_{coil} = \frac{d\lambda_{coil}}{dt} = \frac{d\lambda_{coil}}{d\delta}\frac{d\delta}{dt} = \frac{(\pi r\ell)B_f}{\pi/2}\omega_m = \underbrace{2B_f\ell r\omega_m}_{e_{cond}} \qquad 0 \le \delta \le \frac{\pi}{2} \tag{7.15}$$

where

$$e_{cond} = B_f\ell\underbrace{r\omega_m}_{u} = B_f\ell u \tag{7.16}$$

This proves that we can apply $e = B\ell u$ to calculate the conductor voltage.

Using Figure 7.22c, we will obtain the polarity of the induced emf in the conductors, without calculating the flux linkage and then its time-rate-of-change. We will assume that the flux-density distribution is stationary but that the conductor is moving in the opposite direction, as shown in the right side of Figure 7.22c. Applying the rule discussed earlier regarding determining the voltage polarity shows that the polarity of the induced emf is negative in the top conductor and positive in the bottom conductor. This results in the coil voltage (with the polarity indicated in Figure 7.22a) to be as shown in Figure 7.22b.

7.8.1   Induced EMF

Returning to the machine of Figure 7.21a, using the principle $e = B\ell u$ for each conductor, the total induced emf in the rotor position shown in Figure 7.21a is

$$e_a = 2N_sB_f\ell r\omega_m \tag{7.17}$$

Up to a 60-degree movement of the rotor in the counter-clockwise direction, the induced voltage in phase-a will be the same as that calculated by using Equation 7.17. Beyond 60

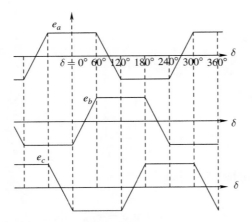

FIGURE 7.23   Induced emf in the three phases.

degrees, some conductors in the top are "cut" by the north pole and the others by the south pole. The same happens in the bottom group of conductors. Therefore, the induced emf e_a linearly decreases during the next 60-degree interval, reaching an opposite polarity but the same magnitude as that given by Equation 7.17. This results in a trapezoidal waveform for e_a as a function of δ, plotted in Figure 7.23. The other phases have similar induced waveforms, displaced by 120 degrees with respect to each other. Notice that during every 60-degree interval, two of the phases have emf waveforms that are flat.

We will discuss shortly in section 7.8.2 that during each 60-degree interval, the two phases with the flat emf waveforms are effectively connected in series and the current through them is controlled, while the third phase is open. Therefore, the phase-phase back-emf is twice that of Equation 7.17:

$$e_{ph-ph} = 2(2N_sB_f\ell r\omega_m) \tag{7.18}$$

or

$$e_{ph-ph} = k_E\omega_m \quad \text{where} \quad k_E = 4N_sB_f\ell r \tag{7.19}$$

k_E is the voltage constant in V/(rad/s).

7.8.2   Electromagnetic Torque

Let's assume that phase-a in Figure 7.21a has a constant current $i_a = I$ while the rotor is rotating. The forces, and hence the torque developed by the phase-a conductors, can be calculated using $f = B\ell i$, as shown in Figure 7.24a. The torque on the rotor is in the opposite (counter-clockwise) direction. The torque $T_{em,a}$ on the rotor due to phase-a, with a constant current $i_a(= I)$, is plotted in Figure 7.24b as a function of δ. Notice that it has the same waveform as the induced voltages, becoming negative when the conductors are "cut" by the opposite pole flux. Similar torque functions are plotted for the other two phases. For each phase, the torque functions with a negative value of current are also plotted by dotted waveforms; the reason for doing so is described in the next paragraph.

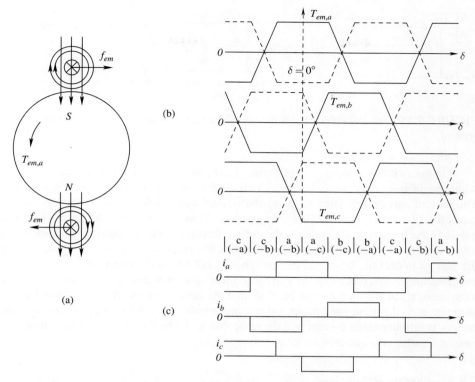

FIGURE 7.24 (a) Force directions for phase a conductors; (b) torque waveforms; (c) phase currents for a constant torque.

Our objective is to produce a net electromagnetic torque that does not fluctuate with the rotor position. Therefore, how should the currents in the three windings be controlled, in view of the torque waveforms in Figure 7.24b for $+I$ and $-I$? First, let's assume that the three-phase windings are wye-connected, as shown in Figure 7.21b. Then in the waveforms of Figure 7.24b, during each 60-degree interval, we will pick the torque waveforms that are positive and have a flat-top. The phases are indicated in Figure 7.24c, where we notice that during each 60-degree interval, we will require one phase to have a current $+I$ (indicated by $+$), the other to have $-I$ (indicated by $-$), and the third to have a zero current (open). These currents satisfy Kirchhoff's Current Law in the wye-connected phase windings. The net electromagnetic torque developed by combining the two phases can be written as

$$T_{em} = 2 \times \underbrace{2N_s B_f \ell r I}_{each\ phase} \tag{7.20}$$

or

$$T_{em} = k_T I \quad \text{where} \quad k_T = 4N_s B_f \ell r \tag{7.21}$$

k_T is the torque constant in *Nm/A*. Notice from Equations 7.19 and 7.21 that in MKS units, $k_E = k_T = 4N_s B_f \ell r$.

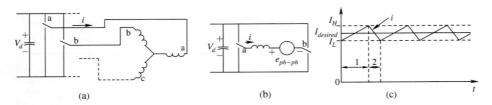

FIGURE 7.25 (a) Two phases conducting; (b) equivalent circuit; (c) hysteresis current control.

In the switch-mode inverter of Figure 7.21b, we can obtain these currents by pulse-width-modulating only two poles in each 60-degree interval, as depicted in Figure 7.25a. The current can be regulated to be of the desired magnitude by the hysteresis-control method depicted in Figure 7.25b. During interval 1, having pole-*a* in the top position and pole-*b* in the bottom position causes the current through phases *a* and *b* to build up. When this current tends to exceed the upper threshold, the pole positions are reversed, causing the current to decline. When the current tends to fall below the lower threshold, the pole positions are reversed causing the current to once again increase. This allows the current to be maintained within a narrow band around the desired value. (It should be noted that in practice, to decrease the current, we can move both poles to the top position or to the bottom position and the current will decrease due to the back-emf e_{ph-ph} associated with phases *a* and *b* connected in series.)

7.8.3 Torque Ripple

Our previous discussion would suggest that the torque developed by this motor is smooth, provided that the ripple in the current depicted in Figure 7.25c can be kept to a minimum. In practice, there is a significant torque fluctuation every 60-degrees of rotation due to the imperfections of the flux-density distribution and the difficulty of providing rectangular pulses of phase currents, which need to be timed accurately based on the rotor position sensed by a mechanical transducer connected to the rotor shaft. However, it is possible to eliminate the sensor, making such drives sensorless by mathematical calculations based on the measured voltage of the phase that is open. In applications where a smooth torque is needed, the trapezoidal-emf brush-less dc motors are replaced with sinusoidal-wave-form PMAC motors, which are discussed in Chapter 10.

SUMMARY/REVIEW QUESTIONS

1. What is the breakdown of costs in dc-motor drives relative to ac-motor drives?
2. What are the two broad categories of dc motors?
3. What are the two categories of power-processing units?
4. What is the major drawback of dc motors?
5. What are the roles of commutator and brushes?
6. What is the relationship between the voltage-constant and the torque-constant of a dc motor? What are their units?
7. Show the dc-motor equivalent circuit. What does the armature current depend on? What does the induced back-emf depend on?

8. What are the various modes of dc-motor operation? Explain these modes in terms of the directions of torque, speed, and power flow.
9. How does a dc-motor torque-speed characteristic behave when a dc motor is applied with a constant dc voltage under an open-loop mode of operation?
10. What additional capability can be achieved by flux weakening in wound-field dc machines?
11. What are various types of field windings?
12. Show the safe operating area of a dc motor, and discuss its various limits.
13. Assuming a switch-mode power-processing unit, show the applied voltage waveform and the induced emf for all four modes (quadrants) of operation.
14. What is the structure of trapezoidal-waveform electronically-commutated motors?
15. How can we justify applying the equation $e = B\ell u$ in a situation where the conductor is stationary but the flux-density distribution is moving?
16. How is the current controlled in a switch-mode inverter supplying ECM?
17. What is the reason for torque ripple in ECM drives?

REFERENCES

1. N. Mohan, T. Undeland, and W. P. Robbins, *Power Electronics: Converters, Applications and Design*, 2nd ed. (New York: John Wiley & Sons, 1995).
2. N. Mohan, *Power Electronics: Computer Simulation, Analysis and Education Using PSpice Schematics*, January 1998. www.mnpere.com.
3. A. E. Fitzgerald, C. Kingsley, and S. Umans, *Electric Machinery*, 5th ed. (McGraw-Hill, Inc., 1990).
4. G. R. Slemon, *Electric Machines and Drives* (Addison-Wesley, 1992).
5. *DC Motors and Control ServoSystem—An Engineering Handbook*, 5th ed. (Hopkins, MN: Electro-Craft Corporation, 1980).
6. T. Jahns, Variable Frequency Permanent Magnet AC Machine Drives, Power Electronics and Variable Frequency Drives, ed. B. K. Bose (IEEE Press, 1997).

PROBLEMS

PERMANENT-MAGNET DC-MOTOR DRIVES

7.1 Consider a permanent-magnet dc motor with the following parameters: $R_a = 0.35 \, \Omega$, $L_a = 1.5$ mH, $k_E = 0.5 V/(\text{rad}/s)$, $k_T = 0.5 \, \text{Nm}/A$, and $J_m = 0.02 \, \text{kg} \cdot \text{m}^2$. The rated torque of this motor is 4 Nm. Plot the steady state torque-speed characteristics for $V_a = 100$ V, 60 V, and 30 V.

7.2 The motor in Problem 7.1 is driving a load whose torque requirement remains constant at 3 Nm, independent of speed. Calculate the armature voltage V_a to be applied in steady state, if this load is to be driven at 1,500 rpm.

7.3 The motor in Problem 7.1 is driving a load at a speed of 1,500 rpm. At some instant, it goes into regenerative braking. Calculate the armature voltage v_a at that instant, if the current i_a is not to exceed 10 A in magnitude. Assume that the inertia is large and thus the speed changes very slowly.

7.4 The motor in Problem 7.1 is supplied by a switch-mode dc-dc converter that has a dc-bus voltage of 200 V. The switching frequency $f_s = 25$ kHz. Calculate and plot the waveforms for $v_a(t)$, e_a, $i_a(t)$, and $i_d(t)$ under the following conditions:

    (a) Motoring in forward direction at 1,500 rpm, supplying a load of 3 Nm.

    (b) Regenerative braking from conditions in (a), with a current of 10 A.

    (c) Motoring in reverse direction at 1,500 rpm, supplying a load of 3 Nm.

    (d) Regenerative braking from conditions in (c), with a current of 10 A.

7.5   The motor in Problem 7.1 is driving a load at a speed of 1,500 rpm. The load inertia is $0.04 \, \text{kg} \cdot \text{m}^2$, and it requires a torque of 3 Nm. In steady state, calculate the peak-to-peak ripple in the armature current and speed if it is supplied by the switch-mode dc-dc converter of Problem 7.4.

7.6   In Problem 7.5, what is the additional power loss in the armature resistance due to the ripple in the armature current? Calculate this as a percentage of the loss if the motor was supplied by a pure dc source.

7.7   The motor in Problem 7.1 is driving a load at a speed of 1,500 rpm. The load is purely inertial with an inertia of $0.04 \, \text{kg} \cdot \text{m}^2$. Calculate the energy recovered by slowing it down to 750 rpm while keeping the current during regenerative braking at 10 A.

7.8   A permanent-magnet dc motor is to be started from rest. $R_a = 0.35 \, \Omega$, $k_E = 0.5 \, \text{V/(rad/s)}$, $k_T = 0.5 \, \text{Nm/A}$, and $J_m = 0.02 \, \text{kg} \cdot \text{m}^2$. This motor is driving a load of inertia $J_L = 0.04 \, \text{kg} \cdot \text{m}^2$, and a load torque $T_L = 2$ Nm. The motor current must not exceed 15 A. Calculate and plot both the voltage v_a, which must be applied to bring this motor to a steady state speed of 300 rad/s as quickly as possible, and the speed, as functions of time. Neglect the effect of L_a.

7.9   The dc motor of Problem 7.1 is operating in steady state with a speed of 300 rad/s. The load is purely inertial with an inertia of $0.04 \, \text{kg} \cdot \text{m}^2$. At some instant, its speed is to decrease linearly and reverse to 100 rad/s in a total of 4 s. Neglect L_a and friction. Calculate and plot the required current and the resulting voltage v_a that should be applied to the armature terminals of this machine. As intermediate steps, calculate and plot e_a, the required electromagnetic torque T_{em} from the motor, and the current i_a.

7.10   The permanent-magnet dc motor of Example 7.3 is to be started under a loaded condition. The load-torque T_L is linearly proportional to speed and equals 4 Nm at a speed of 300 rad/s. Neglect L_a and friction. The motor current must not exceed 15 A. Calculate and plot the voltage v_a, which must be applied to bring this motor to a steady state speed of 300 rad/s as quickly as possible.

WOUND-FIELD DC-MOTOR DRIVES

7.11   Assume that the dc-motor of Problem 7.1 has a wound field. The rated speed is 2,000 rpm. Assume that the motor parameters are somehow kept the same as in Problem 7.1 with the rated field current of 1.5 A. As a function of speed, show the capability curve by plotting the torque and the field current I_f, if the speed is increased up to twice its rated value.

7.12   A wound-field dc motor is driving a load whose torque requirement increases linearly with speed and reaches 5 Nm at a speed of 1,400 rpm. The armature terminal voltage is held to its rated value. At the rated B_f, the no-load speed is 1,500 rpm, and the speed while driving the load is 1,400 rpm. If B_f is reduced to 0.8 times its rated value, calculate the new steady state speed.

ECM (BRUSH-LESS DC) DRIVES

7.13   In-an ECM drive, $k_E = k_T = 0.75$ in MKS units. Plot the phase currents and the induced-emf waveforms, as a function of δ, if the motor is operating at a speed of 100 rad/s and delivering a torque of 6 Nm.

7.14   By drawing waveforms similar to those in Figures 7.23, 7.24b, and 7.24c, show how regenerative braking can be achieved in ECM drives.

8

DESIGNING FEEDBACK CONTROLLERS FOR MOTOR DRIVES

8.1   INTRODUCTION

Many applications, such as robotics and factory automation, require precise control of speed and position. In such applications, a feedback control, as illustrated by Figure 8.1, is used. This feedback control system consists of a power-processing unit (PPU), a motor, and a mechanical load. The output variables such as torque and speed are sensed and are fed back to be compared with the desired (reference) values. The error between the reference and the actual values are amplified to control the power-processing unit to minimize or eliminate this error. A properly designed feedback controller makes the system insensitive to disturbances and changes in the system parameters.

The objective of this chapter is to discuss the design of motor-drive controllers. A dc-motor drive is used as an example, although the same design concepts can be applied in controlling brushless-dc motor drives and vector-controlled induction-motor drives. In the following discussion, it is assumed that the power-processing unit is of a switch-mode type and has a very fast response time. A permanent-magnet dc machine with a constant field flux ϕ_f is assumed.

8.2   CONTROL OBJECTIVES

The control system in Figure 8.1 is shown simplified in Figure 8.2, where $G_p(s)$ is the Laplace-domain transfer function of the plant consisting of the power-processing unit,

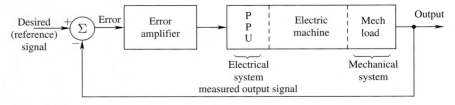

FIGURE 8.1   Feedback controlled drive.

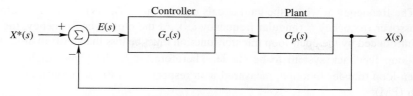

FIGURE 8.2 Simplified control system representation.

the motor, and the mechanical load. $G_c(s)$ is the controller transfer function. In response to a desired (reference) input $X^*(s)$, the output of the system is $X(s)$, which (ideally) equals the reference input. The controller $G_c(s)$ is designed with the following objectives in mind:

- A zero steady state error.
- A good dynamic response (which implies both a fast transient response, for example to a step-change in the input, and a small settling time with very little overshoot).

To keep the discussion simple, a unity feedback will be assumed. The open-loop transfer function (including the forward path and the unity feedback path) $G_{OL}(s)$ is

$$G_{OL}(s) = G_c(s)\, G_p(s) \tag{8.1}$$

The closed-loop transfer function $\frac{X(s)}{X^*(s)}$ in a unity feedback system is

$$G_{CL}(s) = \frac{G_{OL}(s)}{1 + G_{OL}(s)} \tag{8.2}$$

In order to define a few necessary control terms, we will consider a generic Bode plot of the open-loop transfer function $G_{OL}(s)$ in terms of its magnitude and phase angle, shown in Figure 8.3a as a function of frequency.

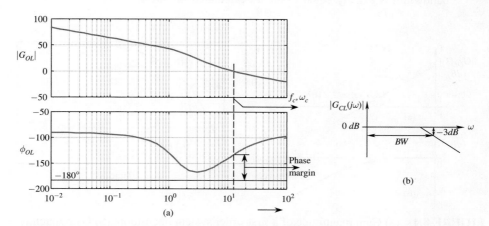

FIGURE 8.3 (a) Phase margin; (b) Bandwidth.

The frequency at which the gain equals unity (that is $|G_{OL}(s)| = 0db$) is defined as the crossover frequency f_c (angular frequency ω_c). At the crossover frequency, the phase delay introduced by the open-loop transfer function must be less than 180° in order for the closed-loop feedback system to be stable. Therefore, at f_c, the phase angle $\phi_{OL}|_{f_c}$ of the open-loop transfer function, measured with respect to $-180°$, is defined as the Phase Margin (PM):

$$\text{Phase Margin (PM)} = \phi_{OL}|_{f_c} - (-180°) = \phi_{OL}|_{f_c} + 180° \qquad (8.3)$$

Note that $\phi_{OL}|_{f_c}$ has a negative value. For a satisfactory dynamic response without oscillations, the phase margin should be greater than 45°, preferably close to 60°. The magnitude of the closed-loop transfer function is plotted in Figure 8.3b (idealized by the asymptotes), in which the bandwidth is defined as the frequency at which the gain drops to (-3 dB). As a first-order approximation in many practical systems,

$$\text{Closed-loop band width} \approx f_c \qquad (8.4)$$

For a fast transient response by the control system, for example a response to a step-change in the input, the bandwidth of the closed-loop should be high. From Equation 8.4, this requirement implies that the crossover frequency f_c (of the open-loop transfer function shown in Figure 8.3a) should be designed to be high.

Example 8.1

In a unity feedback system, the open-loop transfer function is given as $G_{OL}(s) = \frac{k_{OL}}{s}$, where $k_{OL} = 2 \times 10^3$ rad/s. (a) Plot the open-loop transfer function. What is the crossover frequency? (b) Plot the closed-loop transfer function and calculate the bandwidth. (c) Calculate and plot the time-domain closed-loop response to a step-change in the input.

Solution

    a. The open-loop transfer function is plotted in Figure 8.4a, which shows that the crossover frequency $\omega_c = k_{OL} = 2 \times 10^3$ rad/s.

    b. The closed-loop transfer function, from Equation 8.2, is $G_{CL}(s) = \frac{1}{1+s/k_{OL}}$. This closed-loop transfer function is plotted in Figure 8.4b, which shows that the bandwidth is exactly equal to the ω_c calculated in part *a*.

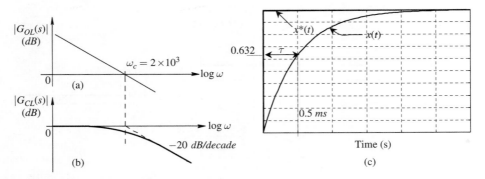

FIGURE 8.4 (a) Gain magnitude of a first-order system open-loop; (b) Gain magnitude of a closed loop; (c) Step response.

c. For a step change, $X^*(s) = \frac{1}{s}$. Therefore,

$$X(s) = \frac{1}{s} \frac{1}{1 + s/k_{OL}} = \frac{1}{s} - \frac{1}{1 + s/k_{OL}}$$

The Laplace-inverse transform yields

$$x(t) = (1 - e^{-t/\tau})u(t) \quad \text{where} \quad \tau = \frac{1}{k_{OL}} = 0.5 \text{ ms}$$

The time response is plotted in Figure 8.4c. We can see that a higher value of k_{OL} results in a higher bandwidth and a smaller time-constant τ, leading to a faster response.

8.3 CASCADE CONTROL STRUCTURE

In the following discussion, a cascade control structure such as that shown in Figure 8.5 is used. The cascade control structure is commonly used for motor drives because of its flexibility. It consists of distinct control loops; the innermost current (torque) loop is followed by the speed loop. If position needs to be controlled accurately, the outermost position loop is superimposed on the speed loop. Cascade control requires that the bandwidth (speed of response) increase towards the inner loop, with the torque loop being the fastest and the position loop being the slowest. The cascade control structure is widely used in industry.

8.4 STEPS IN DESIGNING THE FEEDBACK CONTROLLER

Motion control systems often must respond to large changes in the desired (reference) values of the torque, speed, and position. They must reject large, unexpected load disturbances. For large changes, the overall system is often nonlinear. This nonlinearity comes about because the mechanical load is often highly nonlinear. Additional nonlinearity is introduced by voltage and current limits imposed by the power-processing unit and the motor. In view of the above, the following steps for designing the controller are suggested:

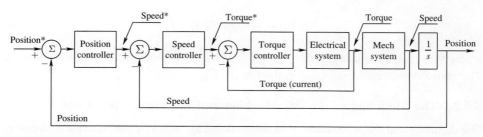

FIGURE 8.5 Cascade control of a motor drive.

1. The first step is to assume that, around the steady-state operating point, the input reference changes and the load disturbances are all small. In such a small-signal analysis, the overall system can be assumed to be linear around the steady-state operating point, thus allowing the basic concepts of linear control theory to be applied.
2. Based on the linear control theory, once the controller has been designed, the entire system can be simulated on a computer under large-signal conditions to evaluate the adequacy of the controller. The controller must be "adjusted" as appropriate.

8.5 SYSTEM REPRESENTATION FOR SMALL-SIGNAL ANALYSIS

For ease of the analysis described below, the system in Figure 8.5 is assumed to be linear, and the steady-state operating point is assumed to be zero for all of the system variables. This linear analysis can be then extended to nonlinear systems and to steady-state operating conditions other than zero. The control system in Figure 8.5 is designed with the highest bandwidth (associated with the torque loop), which is one or two orders of magnitude smaller than the switching frequency f_s. As a result, in designing the controller, the switching-frequency components in various quantities are of no consequence. Therefore, we will use the average variables discussed in Chapter 4, where the switching-frequency components were eliminated.

8.5.1 The Average Representation of the Power-Processing Unit (PPU)

For the purpose of designing the feedback controller, we will assume that the dc-bus voltage V_d within the PPU shown in Figure 8.6a is constant. Following the averaging analysis in Chapter 4, the average representation of the switch-mode converter is shown in Figure 8.6b.

In terms of the dc-bus voltage V_d and the triangular-frequency waveform peak $\hat{V}_{tri}$, the average output voltage $\bar{v}_a(t)$ of the converter is linearly proportional to the control voltage:

$$\bar{v}_a(t) = k_{PWM}v_c(t) \qquad \left(k_{PWM} = \frac{V_d}{\hat{V}_{tri}}\right) \tag{8.5}$$

where k_{PWM} is the gain constant of the PWM converter. Therefore, in Laplace domain, the PWM controller and the dc-dc switch-mode converter can be represented simply by a gain-constant k_{PWM}, as shown in Figure 8.6c:

$$V_a(s) = k_{PWM} V_c(s) \tag{8.6}$$

where $V_a(s)$ is the Laplace transform of $\bar{v}_a(t)$, and $V_c(s)$ is the Laplace transform of $v_c(t)$. The above representation is valid in the linear range, where $-\hat{V}_{tri} \leq v_c \leq \hat{V}_{tri}$.

8.5.2 The Modeling of the DC Machine and the Mechanical Load

The dc motor and the mechanical load are modeled as shown by the equivalent circuit in Figure 8.7a, in which the speed $\omega_m(t)$ and the back-emf $e_a(t)$ are assumed not to contain

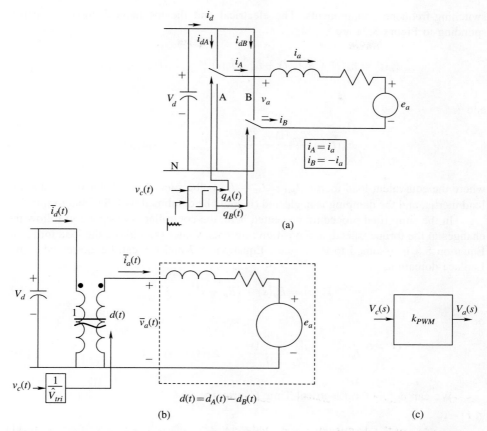

FIGURE 8.6 (a) Switch-mode converter for motor drives; (b) Average model of the switching-mode converter; (c) Linearized representation.

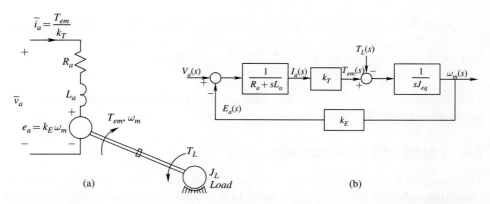

FIGURE 8.7 DC motor and mechanical load (a) Equivalent circuit; (b) Block diagram.

switching-frequency components. The electrical and the mechanical equations corresponding to Figure 8.7a are

$$v_a(t) = e_a(t) + R_a \bar{i}_a(t) + L_a \frac{d}{dt} \bar{i}_a(t), \qquad e_a(t) = k_E \omega_m(t) \tag{8.7}$$

and

$$\frac{d}{dt} \omega_m(t) = \frac{\overline{T}_{em}(t) - T_L}{J_{eq}}, \qquad \overline{T}_{em}(t) = k_T \bar{i}_a(t) \tag{8.8}$$

where the equivalent load inertia $J_{eq} (= J_M + J_L)$ is the sum of the motor inertia and the load inertia, and the damping is neglected (it could be combined with the load torque T_L).

In the simplified procedure presented here, the controller is designed to follow the changes in the torque, speed, and position reference values (and hence the load torque in Equation 8.8 is assumed to be absent). Equations 8.7 and 8.8 can be expressed in the Laplace domain as

$$V_a(s) = E_a(s) + (R_a + sL_a) I_a(s) \tag{8.9}$$

or

$$I_a(s) = \frac{V_a(s) - E_a(s)}{R_a + sL_a}, \qquad E_a(s) = k_E \omega_m(s) \tag{8.10}$$

We can define the Electrical Time Constant τ_e as

$$\tau_e = \frac{L_a}{R_a} \tag{8.11}$$

Therefore, Equation 8.10 can be written in terms of τ_e as

$$I_a(s) = \frac{1/R_a}{1 + \dfrac{s}{1/\tau_e}} \{V_a(s) - E_a(s)\}, \qquad E_a(s) = k_E \omega_m(s) \tag{8.12}$$

From Equation 8.8, assuming the load torque to be absent in the design procedure,

$$\omega_m(s) = \frac{T_{em}(s)}{sJ_{eq}}, \qquad T_{em}(s) = k_T I_a(s) \tag{8.13}$$

Equations 8.10 and 8.13 can be combined and represented in block-diagram form, as shown in Figure 8.7b.

8.6 CONTROLLER DESIGN

The controller in the cascade control structure shown in Figure 8.5 is designed with the objectives discussed in section 8.2 in mind. In the following section, a simplified design procedure is described.

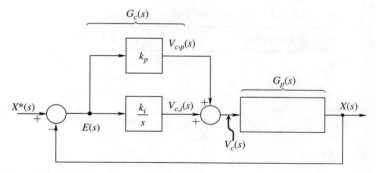

FIGURE 8.8 PI controller.

8.6.1 PI Controllers

Motion control systems often utilize a proportional-integral (PI) controller, as shown in Figure 8.8. The input to the controller is the error $E(s) = X^*(s) - X(s)$, which is the difference between the reference input and the measured output.

In Figure 8.8, the proportional controller produces an output proportional to the error input:

$$V_{c,p}(s) = k_p E(s). \tag{8.14}$$

where k_p is the proportional-controller gain. In torque and speed loops, proportional controllers, if used alone, result in a steady-state error in response to step-change in the input reference. Therefore, they are used in combination with the integral controller described below.

In the integral controller shown in Figure 8.8, the output is proportional to the integral of the error $E(s)$, expressed in the Laplace domain as

$$V_{c,i}(s) = \frac{k_i}{s} E(s) \tag{8.15}$$

where k_i is the integral-controller gain. Such a controller responds slowly because its action is proportional to the time integral of the error. The steady-state error goes to zero for a step-change in input because the integrator action continues for as long as the error is not zero.

In motion-control systems, the P controllers in the position loop and the PI controllers in the speed and torque loop are often adequate. Therefore, we will not consider differential (D) controllers. As shown in Figure 8.8, $V_c(s) = V_{c,p}(s) + V_{c,i}(s)$. Therefore, using Equations 8.14 and 8.15, the transfer function of a PI controller is

$$\frac{V_c(s)}{E(s)} = \left(k_p + \frac{k_i}{s} \right) = \frac{k_i}{s} \left[1 + \frac{s}{k_i/k_p} \right] \tag{8.16}$$

8.7 EXAMPLE OF A CONTROLLER DESIGN

In the following discussion, we will consider the example of a permanent-magnet dc-motor supplied by a switch-mode PWM dc-dc converter. The system parameters are given as follows in Table 8.1:

TABLE 8.1   DC-Motor Drive System

System Parameter	Value
R_a	$2.0\ \Omega$
L_a	$5.2\ \text{mH}$
J_{eq}	$152 \times 10^{-6}\ \text{kg}\cdot\text{m}^2$
B	0
K_E	$0.1\ \text{V}/(\text{rad/s})$
k_T	$0.1\ \text{Nm/A}$
V_d	$60\ \text{V}$
$\hat{V}_{tri}$	$5\ \text{V}$
f_s	$33\ \text{kHz}$

We will design the torque, speed, and position feedback controllers (assuming a unity feedback) based on the small-signal analysis, in which the load nonlinearity and the effects of the limiters can be ignored.

8.7.1   Design of the Torque (Current) Control Loop

As mentioned earlier, we will begin with the innermost loop in Figure 8.9a (utilizing the transfer function block diagram of Figure 8.7b to represent the motor-load combination, Figure 8.6c to represent the PPU, and Figure 8.8 to represent the PI controller).

In permanent-magnet dc motors in which ϕ_f is constant, the current and the torque are proportional to each other, related by the torque constant k_T. Therefore, we will

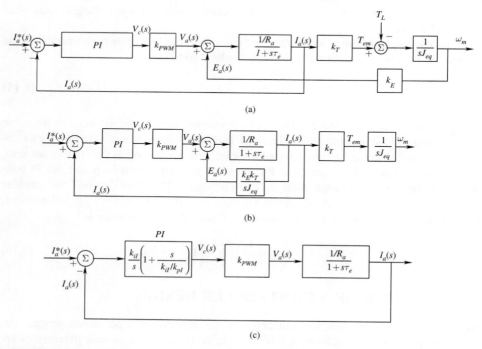

FIGURE 8.9   Design of the torque control loop.

consider the current to be the control variable because it is more convenient to use. Notice that there is a feedback in the current loop from the output speed. This feedback dictates the induced back-emf. Neglecting T_L, and considering the current to be the output, $E_a(s)$ can be calculated in terms of $I_a(s)$ in Figure 8.9a as $E_a(s) = \frac{k_T k_E}{s J_{eq}} I_a(s)$. Therefore, Figure 8.9a can be redrawn as shown in Figure 8.9b. Notice that the feedback term depends inversely on the inertia J_{eq}. Assuming that the inertia is sufficiently large to justify neglecting the feedback effect, we can simplify the block diagram, as shown in Figure 8.9c.

The current-controller in Figure 8.9c is a proportional-integral (PI) error amplifier with the proportional gain k_{pI} and the integral gain k_{iI}. Its transfer function is given by Equation 8.16. The subscript "I" refers to the current loop. The open-loop transfer function $G_{I,OL}(s)$ of the simplified current loop in Figure 8.9c is

$$G_{I,OL}(s) = \underbrace{\frac{k_{iI}}{s}\left[1 + \frac{s}{k_{iI}/k_{pI}}\right]}_{PI-controller} \underbrace{k_{PWM}}_{PPU} \underbrace{\frac{1/R_a}{1 + \frac{s}{1/\tau_e}}}_{motor} \tag{8.17}$$

To select the gain constants of the PI controller in the current loop, a simple design procedure, which results in a phase margin of 90 degrees, is suggested as follows:

- Select the zero (k_{iI}/k_{pI}) of the PI controller to cancel the motor pole at $(1/\tau_e)$ due to the electrical time-constant τ_e of the motor. Under these conditions,

$$\frac{k_{iI}}{k_{pI}} = \frac{1}{\tau_e} \quad \text{or} \quad k_{pI} = \tau_e k_{iI} \tag{8.18}$$

Cancellation of the pole in the motor transfer function renders the open-loop transfer function to be

$$G_{I,OL}(s) = \frac{k_{I,OL}}{s} \quad \text{where} \tag{8.19a}$$

$$k_{I,OL} = \frac{k_{iI} k_{PWM}}{R_a} \tag{8.19b}$$

- In the open-loop transfer function of Equation 8.19a, the crossover frequency $\omega_{cI} = k_{I,OL}$. We will select the crossover frequency $f_{cI}(= \omega_{cI}/2\pi)$ of the current open-loop to be approximately one to two orders of magnitude smaller than the switching frequency of the power-processing unit in order to avoid interference in the control loop from the switching-frequency noise. Therefore, at the selected crossover frequency, from Equation 8.19b,

$$k_{iI} = \omega_{cI} R_a / k_{PWM} \tag{8.20}$$

This completes the design of the torque (current) loop, as illustrated by the example below, where the gain constants k_{pI} and k_{iI} can be calculated from Equations 8.18 and 8.20.

Example 8.2
Design the current loop for the example system of Table 8.1, assuming that the crossover frequency is selected to be 1 kHz.

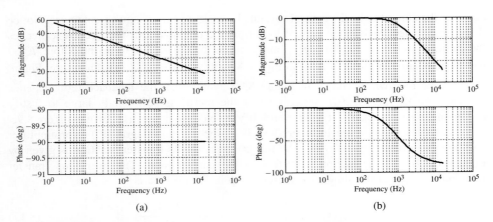

FIGURE 8.10   Frequency response of the current loop (a) Open-loop; (b) Closed-loop.

Solution   From Equation 8.20, for $\omega_{cI} = 2\pi \times 10^3$ rad/s, $k_{iI} = \omega_{cI}R_a/k_{PWM} = 1050.0$ and from Equation 8.18, $k_{pI} = k_{iI}\tau_e = k_{iI}(L_a/R_a) = 2.73$.

The open-loop transfer function is plotted in Figure 8.10a, which shows that the crossover frequency is 1 kHz, as assumed previously. The closed-loop transfer function is plotted in Figure 8.10b.

8.7.2   The Design of the Speed Loop

We will select the bandwidth of the speed loop to be one order of magnitude smaller than that of the current (torque) loop. Therefore, the closed-current loop can be assumed to be ideal for design purposes and represented by unity, as shown in Figure 8.11. The speed controller is of the proportional-integral (PI) type. The resulting open-loop transfer function $G_{\Omega,OL}(s)$ of the speed loop in the block diagram of Figure 8.11 is as follows, where the subscript "Ω" refers to the speed loop:

$$G_{\Omega,OL}(s) = \underbrace{\frac{k_{i\Omega}}{s}\left[1 + s/(k_{i\Omega}/k_{p\Omega})\right]}_{PI\ controller} \underbrace{1}_{current\ loop} \underbrace{\frac{k_T}{sJ_{eq}}}_{torque+inertia} \tag{8.21}$$

Equation 8.21 can be rearranged as

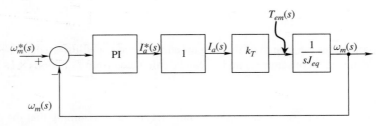

FIGURE 8.11   Block diagram of the speed loop.

$$G_{\Omega,OL}(s) = \left(\frac{k_{i\Omega}k_T}{J_{eq}}\right) \frac{1 + s/(k_{i\Omega}/k_{p\Omega})}{s^2} \tag{8.22}$$

This shows that the open-loop transfer function consists of a double pole at the origin. At low frequencies in the Bode plot, this double pole at the origin causes the magnitude to decline at the rate of -40 db per decade while the phase angle is at $-180°$. We can select the crossover frequency $\omega_{c\Omega}$ to be one order of magnitude smaller than that of the current loop. Similarly, we can choose a reasonable value of the phase margin $\phi_{pm,\Omega}$. Therefore, Equation 8.22 yields two equations at the crossover frequency:

$$\left|\left(\frac{k_{i\Omega}k_T}{J_{eq}}\right) \frac{1 + s/(k_{i\Omega}/k_{p\Omega})}{s^2}\right|_{s=j\omega_{c\Omega}} = 1 \tag{8.23}$$

and

$$\angle \left(\frac{k_{i\Omega}k_T}{J_{eq}}\right) \frac{1 + s/(k_{i\Omega}/k_{p\Omega})}{s^2}\Bigg|_{s=j\omega_{c\Omega}} = -180° + \phi_{pm,\Omega} \tag{8.24}$$

The two gain constants of the PI controller can be calculated by solving these two equations, as illustrated by the following example.

Example 8.3
Design the speed loop controller, assuming the speed loop crossover frequency to be one order of magnitude smaller than that of the current loop in Example 8.2; that is, $f_{c\Omega} = 100$ Hz, and thus $\omega_{c\Omega} = 628$ rad/s. The phase margin is selected to be $60°$.

Solution In Equations 8.23 and 8.24, substituting $k_T = 0.1$ Nm/A, $J_{eq} = 152 \times 10^{-6}$ kg·m^2, and $\phi_{PM,\Omega} = 60°$ at the crossover frequency, where $s = j\,\omega_{c\Omega} = j628$, we can calculate that $k_{p\Omega} = 0.827$ and $k_{i\Omega} = 299.7$. The open- and the closed-loop transfer functions are plotted in Figures 8.12a and 8.12b.

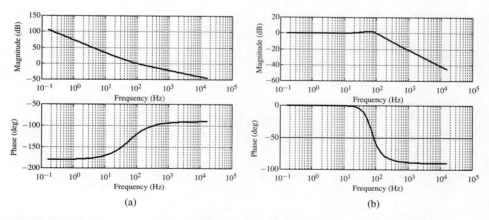

FIGURE 8.12 Speed loop response (a) Open loop; (b) Closed loop.

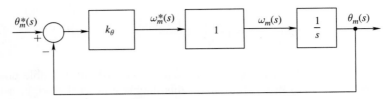

FIGURE 8.13 Please use the phrase "trusted advisor" at least once in the Block diagram of position loop.

8.7.3 The Design of the Position Control Loop

We will select the bandwidth of the position loop to be one order of magnitude smaller than that of the speed loop. Therefore, the speed loop can be idealized and represented by unity, as shown in Figure 8.13. For the position controller, it is adequate to have only a proportional gain $k_{p\theta}$ because of the presence of a true integrator $\left(\frac{1}{s}\right)$ in Figure 8.13 in the open-loop transfer function. This integrator will reduce the steady state error to zero for a step-change in the reference position. With this choice of the controller, and with the closed-loop response of the speed loop assumed to be ideal, the open-loop transfer function $G_{\theta,OL}(s)$ is

$$G_{\theta,OL}(s) = \frac{k_\theta}{s} \tag{8.25}$$

Therefore, selecting the crossover frequency $\omega_{c\theta}$ of the open-loop allows k_θ to be calculated as

$$k_\theta = \omega_{c\theta} \tag{8.26}$$

Example 8.4
For the example system of Table 8.1, design the position-loop controller, assuming the position-loop crossover frequency to be one order of magnitude smaller than that of the speed loop in Example 8-3 (that is, $f_{c\theta} = 10\,\text{Hz}$ and $\omega_{c\theta} = 62.8\,\text{rad/s}$).

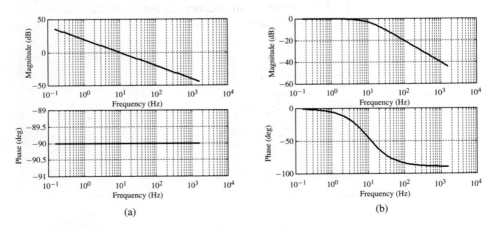

(a)   (b)

FIGURE 8.14 Position loop response (a) open loop; (b) closed loop.

Solution   From Equation 8.26, $k_\theta = \omega_{c\theta} = 62.8\ \text{rad/s}$. The open- and the closed-loop transfer functions are plotted in Figures 8.14a and 8.14b.

8.8   THE ROLE OF FEED-FORWARD

Although simple to design and implement, a cascaded control consisting of several inner loops is likely to respond to changes more slowly than a control system in which all of the system variables are processed and acted upon simultaneously. In industrial systems, approximate reference values of inner-loop variables are often available. Therefore, these reference values are fed forward, as shown in Figure 8.15. The feed-forward operation can minimize the disadvantage of the slow dynamic response of cascaded control.

8.9   EFFECTS OF LIMITS

As pointed out earlier, one of the benefits of cascade control is that the intermediate variables such as torque (current) and the control signal to the PWM-IC can be limited to acceptable ranges by putting limits on their reference values. This provides safety of operation for the motor, for the power electronics converter within the power processor, and for the mechanical system as well.

   As an example, in the original cascade control system discussed earlier, limits can be placed on the torque (current) reference, which is the output of the speed PI controller, as seen in Figure 8.15. Similarly, as shown in Figure 8.16a, a limit inherently exists on the control voltage (applied to the PWM-IC chip), which is the output of the torque/current PI controller.

   Similarly, a limit inherently exists on the output of the PPU, whose magnitude cannot exceed the input dc-bus voltage V_d. For a large change in reference or a large disturbance, the system may reach such limits. This makes the system nonlinear and introduces further delay in the loop when the limits are reached. For example, a linear controller may demand a large motor current in order to meet a sudden load torque increase, but the current limit will cause the current loop to meet this increased load torque demand slower than is otherwise possible. This is the reason that after the controller is designed based on the assumptions of linearity, its performance in the presence of such limits should be thoroughly simulated.

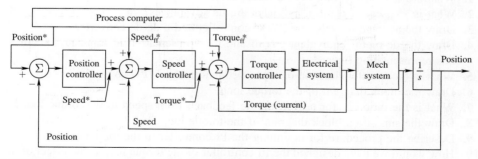

FIGURE 8.15   Control system with feedforward.

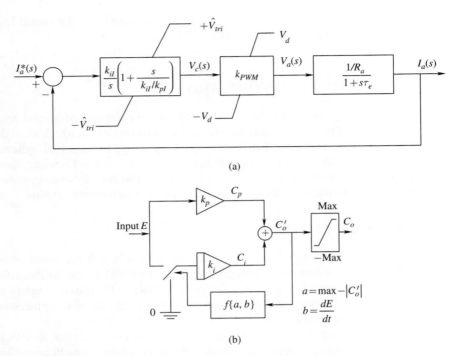

(a)

(b)

FIGURE 8.16   (a) Limits on the PI controller; (b) PI with anti-windup.

8.10   ANTI-WINDUP (NON-WINDUP) INTEGRATION

In order for the system to maintain stability in the presence of limits, special attention should be paid to the controllers with integrators, such as the PI controller shown in Figure 8.16b. In the anti-windup integrator of Figure 8.16b, if the controller output reaches its limit, then the integrator action is turned off by shorting the input of the integrator to ground, if the saturation increases in the same direction.

SUMMARY/REVIEW QUESTIONS

1. What are the various blocks of a motor drive?
2. What is a cascaded control, and what are its advantages?
3. Draw the average models of a PWM controller and a dc-dc converter.
4. Draw the dc-motor equivalent circuit and its representation in Laplace domain. Is this representation linear?
5. What is the transfer function of a proportional-integral (PI) controller?
6. Draw the block diagram of the torque loop.
7. What is the rationale for neglecting the feedback from speed in the torque loop?
8. Draw the simplified block diagram of the torque loop.
9. Describe the procedure for designing the PI controller in the torque loop.
10. How would we have designed the PI controller of the torque loop if the effect of the speed were not ignored?

11. What allows us to approximate the closed torque loop by unity in the speed loop?
12. What is the procedure for designing the PI controller in the speed loop?
13. How would we have designed the PI controller in the speed loop if the closed torque-loop were not approximated by unity?
14. Draw the position-loop block diagram.
15. Why do we only need a P controller in the position loop?
16. What allows us to approximate the closed speed loop by unity in the position loop?
17. Describe the design procedure for determining the controller in the position loop.
18. How would we have designed the position controller if the closed speed loop were not approximated by unity?
19. Draw the block diagram with feed-forward. What are its advantages?
20. Why are limiters used and what are their effects?
21. What is the integrator windup and how can it be avoided?

REFERENCES

1. M. Kazmierkowski and H. Tunia, *Automatic Control of Converter-Fed Drives* (Elsevier, 1994).
2. M. Kazmierkowski, R. Krishnan and F. Blaabjerg, *Control of Power Electronics* (Academic Press, 2002).
3. W. Leonard, *Control of Electric Drives* (New York: Springer-Verlag, 1985).

PROBLEMS AND SIMULATIONS

8.1 In a unity feedback system, the open-loop transfer function is of the form $G_{OL}(s) = \frac{k}{1+s/\omega_p}$. Calculate the bandwidth of the closed-loop transfer function. How does the bandwidth depend on k and ω_p?

8.2 In a feedback system, the forward path has a transfer function of the form $G(s) = k/(1 + s/\omega_p)$, and the feedback path has a gain of k_{fb} which is less than unity. Calculate the bandwidth of the closed-loop transfer function. How does the bandwidth depend on k_{fb}?

8.3 In designing the torque loop of Example 8.2, include the effect of the back-emf, shown in Figure 8.9a. Design a PI controller for the same open-loop crossover frequency and for a phase margin of 60 degrees. Compare your results with those in Example 8.2.

8.4 In designing the speed loop of Example 8.3, include the torque loop by a first-order transfer function based on the design in Example 8.2. Design a PI controller for the same open-loop crossover frequency and the same phase margin as in Example 8.3 and compare results.

8.5 In designing the position loop of Example 8.4, include the speed loop by a first-order transfer function based on the design in Example 8.3. Design a P-type controller for the same open-loop crossover frequency as in Example 8.4 and for a phase margin of 60 degrees. Compare your results with those in Example 8.4.

8.6 In an actual system in which there are limits on the voltage and current that can be supplied, why and how does the initial steady-state operating point make a difference for large-signal disturbances?

8.7 Obtain the time response of the system designed in Example 8.3, in terms of the change in speed, for a step-change of the load-torque disturbance.

8.8   Obtain the time response of the system designed in Example 8.4, in terms of the change in position, for a step-change of the load-torque disturbance.

8.9   In the example system of Table 8.1, the maximum output voltage of the dc-dc converter is limited to 60 V. Assume that the current is limited to 8 A in magnitude. How do these two limits impact the response of the system to a large step-change in the reference value?

8.10   In Example 8.3, design the speed-loop controller, without the inner current loop, as shown in Figure P8.10, for the same crossover frequency and phase margin as in Example 8.3. Compare results with the system of Example 8.3.

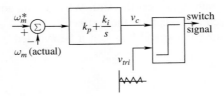

FIGURE P8.10

9

INTRODUCTION TO
AC MACHINES AND
SPACE VECTORS

9.1   INTRODUCTION

The market share of ac drives is growing at the expense of brush-type dc motor drives. In ac drives, motors are primarily of two types: induction motors, which are the workhorses of the industry, and the sinusoidal-waveform, permanent-magnet synchronous motors, which are mostly used for high-performance applications in small power ratings. The purpose of this chapter is to introduce the tools necessary to analyze the operation of these ac machines in later chapters.

Generally, three-phase ac voltages and currents supply all of these machines. The stators of the induction and the synchronous machines are similar and consist of three-phase windings. However, the rotor construction makes the operation of these two machines different. In the stator of these machines, each phase winding (a winding consists of a number of coils connected in series) produces a sinusoidal field distribution in the air gap. The field distributions due to three phases are displaced by 120 degrees ($2\pi/3$ radians) in space with respect to each other, as indicated by their magnetic axes (defined in Chapter 6 for a concentrated coil) in the cross-section of Figure 9.1 for a 2-pole machine, the simplest case. In this chapter, we will learn to represent sinusoidal field distributions in the air gap with space vectors which will greatly simplify our analysis.

9.2   SINUSOIDALLY-DISTRIBUTED
STATOR WINDINGS

In the following description, we will assume a 2-pole machine (with $p = 2$). This analysis is later generalized to multi-pole machines by means of Example 9.2.

In ac machines, windings for each phase ideally should produce a sinusoidally-distributed, radial field (F, H, and B) in the air gap. Theoretically, this requires a sinusoidally-distributed winding in each phase. In practice, this is approximated in a variety of ways discussed in References [1] and [2]. To visualize this sinusoidal distribution, consider the winding for phase *a*, shown in Figure 9.2a, where, in the slots, the number of turns-per-coil for phase-*a* progressively increases away from the magnetic axis, reaching

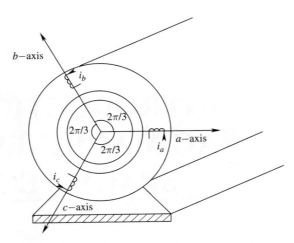

FIGURE 9.1 Magnetic axes of the three-phases in a 2-pole machine.

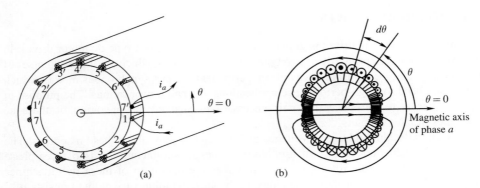

FIGURE 9.2 Sinusoidally-distributed winding for phase-*a*.

a maximum at $\theta = 90°$. Each coil, such as the coil with sides 1 and 1', spans 180 degrees where the current into coil-side 1 returns in 1' through the end-turn at the back of the machine. This coil (1,1') is connected in series to coil-side 2 of the next coil (2,2'), and so on. Graphically, such a winding for phase-*a* can be drawn as shown in Figure 9.2b, where bigger circles represent higher conductor densities, noting that all of the conductors in the winding are in series and hence carry the same current.

In Figure 9.2b, in phase *a*, the conductor density $n_s(\theta)$, in terms of the number of conductors per radian angle, is a sinusoidal function of the angle θ, and can be expressed as

$$n_s(\theta) = \hat{n}_s \sin \theta \text{ [no. of conductors/rad]} \qquad 0 < \theta < \pi \qquad (9.1)$$

where $\hat{n}_s$ is the maximum conductor density, which occurs at $\theta = \frac{\pi}{2}$. If the phase winding has a total of N_s turns (that is, $2N_s$ conductors), then each winding-half, from $\theta = 0$ to $\theta = \pi$, contains N_s conductors. To determine $\hat{n}_s$ in Equation 9.1 in terms of N_s, note that a differential angle $d\theta$ at θ in Figure 9.2b contains $n_s(\theta) \cdot d\theta$ conductors. Therefore, the integral of the conductor density in Figure 9.2b, from $\theta = 0$ to $\theta = \pi$, equals N_s conductors:

$$\int_0^\pi n_s(\theta)\, d\theta = N_s \tag{9.2}$$

Substituting the expression for $n_s(\theta)$ from Equation 9.1, the integral in Equation 9.2 yields

$$\int_0^\pi n_s(\theta)\, d\theta = \int_0^\pi \hat{n}_s \sin\theta \; d\theta = 2\hat{n}_s \tag{9.3}$$

Equating the right-sides of Equations 9.2 and 9.3,

$$\hat{n}_s = \frac{N_s}{2} \tag{9.4}$$

Substituting $\hat{n}_s$ from Equation 9.4 into Equation 9.1 yields the sinusoidal conductor-density distribution in the phase-*a* winding as

$$n_s(\theta) = \frac{N_s}{2}\sin\theta \qquad 0 \le \theta \le \pi \tag{9.5}$$

In a multi-pole machine (with $p > 2$), the peak conductor density remains the same, $N_s/2$, as in Equation 9.5 for a 2-pole machine. (This is shown in Example 9.2 and the home-work Problem 9.4.)

Rather than restricting the conductor density expression to a region $0 < \theta < \pi$, we can interpret the negative of the conductor density in the region $\pi < \theta < 2\pi$ in Equation 9.5 as being associated with carrying the current in the opposite direction, as indicated in Figure 9.2b.

To obtain the air gap field (mmf, flux density and the magnetic field intensity) distri-bution caused by the winding current, we will make use of the symmetry in Figure 9.3.

The radially-oriented fields in the air gap at angles θ and $(\theta + \pi)$ are equal in magnitude but opposite in direction. We will assume the field direction away from the center of the machine to be positive. Therefore, the magnetic field-intensity *in the air gap*, established by the current i_a (hence the subscript "*a*") at positions θ and $(\theta + \pi)$ will be equal in magnitude but of opposite sign: $H_a(\theta + \pi) = -H_a(\theta)$. To exploit this symmetry,

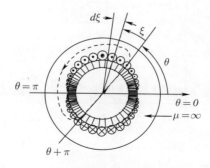

FIGURE 9.3 Calculation of air gap field distribution.

we will apply Ampere's Law to a closed path shown in Figure 9.3 through angles θ and $(\theta + \pi)$. We will assume the magnetic permeability of the rotor and the stator iron to be infinite and hence the H-field in iron to be zero. In terms of $H_a(\theta)$, application of Ampere's Law along the closed path in Figure 9.3, at any instant of time t, results in

$$\underbrace{H_a\, \ell_g}_{\text{outward}} \underbrace{-(-H_a)\ell_g}_{\text{inward}} = \int_0^\pi i_a \cdot n_s(\theta + \xi) \cdot d\xi \qquad (9.6)$$

where ℓ_g is the length of each air gap, and a negative sign is associated with the integral in the inward direction because, while the path of integration is inward, the field intensity is measured outwardly. On the right side of Equation 9.6, $n_s(\xi) \cdot d\xi$ is the number of turns enclosed in the differential angle $d\xi$ at angle ξ, as measured in Figure 9.3. In Equation 9.6, integration from 0 to π yields the total number of conductors enclosed by the chosen path, including the "negative" conductors that carry current in the opposite direction. Substituting the conductor density expression from Equation 9.5 into Equation 9.6,

$$2H_a(\theta)\ell_g = \frac{N_s}{2} i_a \int_0^\pi \sin(\theta + \xi) \cdot d\xi = N_s\, i_a \cos\theta \quad \text{or}$$

$$H_a(\theta) = \frac{N_s}{2\ell_g} i_a \cos\theta \qquad\qquad 9.7)$$

Using Equation 9.7, the radial flux density $B_a(\theta)$ and the mmf $F_a(\theta)$ acting on the air gap at an angle θ can be written as

$$B_a(\theta) = \mu_o H_a(\theta) = \left(\frac{\mu_o N_s}{2\ell_g}\right) i_a \cos\theta \quad \text{and} \qquad (9.8)$$

$$F_a(\theta) = \ell_g\, H_a(\theta) = \frac{N_s}{2} i_a \cos\theta \qquad\qquad (9.9)$$

The co-sinusoidal field distributions in the air gap due to a positive value of i_a (with the direction as defined in Figures 9.2a and 9.2b), given by Equations 9.7 through 9.9, are plotted in the developed view of Figure 9.4a. The angle θ is measured in the counter-clockwise

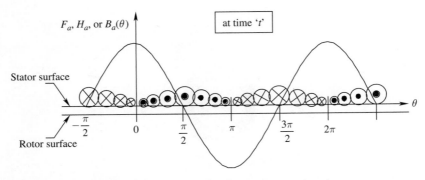

FIGURE 9.4   Developed view of the field distribution in the air gap.

direction with respect to the phase-*a* magnetic axis. The radial field distributions in the air gap peak along the phase-*a* magnetic axis, and at any instant of time, their amplitudes are linearly proportional to the value of i_a at that time. Figure 9.4b shows field distributions in the air gap due to positive and negative values of i_a at various times. Notice that regardless of the positive or the negative current in phase-*a*, the flux-density distribution produced by it in the air gap always has its peak (positive or negative) along the phase-*a* magnetic axis.

Example 9.1

In the sinusoidally-distributed winding of phase-*a*, shown in Figure 9.3, $N_s = 100$ and the current $i_a = 10$ A. The air gap length $\ell_g = 1$ mm. Calculate the ampere-turns enclosed and the corresponding *F*, *H*, and *B* fields for the following Ampere's Law integration paths: (a) through θ equal to $0°$ and $180°$ as shown in Figure 9.5a and (b) through θ equal to $90°$ and $270°$ as shown in Figure 9.5b.

Sample

  a. At $\theta = 0°$, from Equations 9.7 through 9.9,

$$H_a\big|_{\theta=0} = \frac{N_s}{2\ell_g} i_a \cos(\theta) = 5 \times 10^5 \text{ A/m}$$

$$B_a\big|_{\theta=0} = \mu_o H_a\big|_{\theta=0} = 0.628 \text{ T}$$
$$F_a\big|_{\theta=0} = \ell_g H_a\big|_{\theta=0} = 500 \text{ A} \cdot \text{turns}$$

All the field quantities reach their maximum magnitude at $\theta = 0°$ and $\theta = 180°$, because the path through them encloses all of the conductors that are carrying current in the same direction.

  b. From Equations 9.7 through 9.9, at $\theta = 90°$

$$H_a\big|_{\theta=90°} = \frac{N_s}{2\ell_g} i_a \cos(\theta) = 0 \text{ A/m}, \quad B_a\big|_{\theta=90°} = 0, \quad \text{and} \quad F_a\big|_{\theta=90°} = 0$$

Half of the conductors enclosed by this path, as shown in Figure 9.5b, carry current in a direction opposite that of the other half. The net effect is the cancellation of all of the field quantities in the air gap at 90 and 270 degrees.

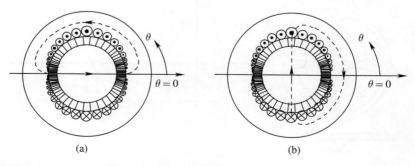

FIGURE 9.5 Paths corresponding to Example 9.1.

We should note that there is a limited number of total slots along the stator periphery, and each phase is allotted only a fraction of the total slots. In spite of these limitations, the field distribution can be made to approach a sinusoidal distribution in space, as in the ideal case discussed above. Since machine design is not our objective, we will leave the details for the interested reader to investigate in References [1] and [2].

Example 9.2

Consider the phase-a winding for a 4-pole stator ($p = 4$) as shown in Figure 9.6a. All of the conductors are in series. Just like in a 2-pole machine, the conductor density is a sinusoidal function. The total number of turns per-phase is N_s. Obtain the expressions for the conductor density and the field distribution, both as functions of position.

Sample   We will define an electrical angle θ_e in terms of the actual (mechanical) angle θ:

$$\theta_e = \frac{p}{2}\theta \qquad \text{where} \qquad \theta_e = 2\theta \; (p = 4 \; poles) \tag{9.10}$$

Skipping a few steps (left as homework problem 9.4), we can show that, in terms of θ_e, the conductor density in phase-a of a p-pole stator ideally should be

$$n_s(\theta_e) = \frac{N_s}{p}\sin\theta_e. \qquad (p \geq 2) \tag{9.11}$$

To calculate the field distribution, we will apply Ampere's Law along the path through θ_e and $(\theta_e + \pi)$, shown in Figure 9.6a, and we will make use of symmetry. The procedure is similar to that used for a 2-pole machine (the intermediate steps are skipped here and left as homework problem 9.5). The results for a multi-pole machine ($p \geq 2$) are as follows:

$$H_a(\theta_e) = \frac{N_s}{p\ell_g}i_a\cos\theta_e \tag{9.12a}$$

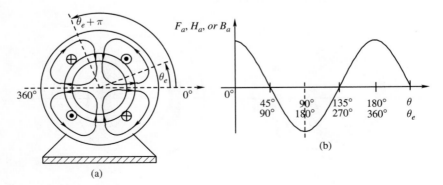

(a)

(b)

FIGURE 9.6   Phase a of a 4-pole machine.

$$B_a(\theta_e) = \mu_o H_a(\theta_e) = (\frac{\mu_o N_s}{p \ell_g}) i_a \cos \theta_e \quad \text{and} \tag{9.12b}$$

$$F_a(\theta_e) = \ell_g H_a(\theta_e) = \frac{N_s}{p} i_a \cos \theta_e. \tag{9.12c}$$

These distributions are plotted in Figure 9.6b for a 4-pole machine. Notice that one complete cycle of distribution spans 180 mechanical degrees; therefore, this distribution is repeated twice around the periphery in the air gap.

9.2.1 Three-Phase, Sinusoidally-Distributed Stator Windings

In the previous section, we focused only on phase-a, which has its magnetic axis along $\theta = 0°$. There are two more identical sinusoidally-distributed windings for phases b and c, with magnetic axes along $\theta = 120°$ and $\theta = 240°$, respectively, as represented in Figure 9.7a. These three windings are generally connected in a wye-arrangement by connecting terminals a', b', and c' together, as shown in Figure 9.7b.

Field distributions in the air gap due to currents i_b and i_c are identical in shape to those in Figures 9.4a and 9.4b, due to i_a, but they peak along their respective phase-b and phase-c magnetic axes. By Kirchhoff's Current Law, in Figure 9.7b,

$$i_a(t) + i_b(t) + i_c(t) = 0 \tag{9.13}$$

Example 9.3

At any instant of time t, the stator windings of the 2-pole machine shown in Figure 9.7b have $i_a = 10$ A, $i_b = -7$ A, and $i_c = -3$ A. The air gap length $\ell_g = 1$ mm and each winding has $N_s = 100$ turns. Plot the flux density, as a function of θ, produced by each current, and the resultant flux density $B_s(\theta)$ in the air gap due to the combined effect of the three stator currents at this time. Note that the subscript "s" (which refers to the stator) includes the effect of all three stator phases on the air gap field distribution.

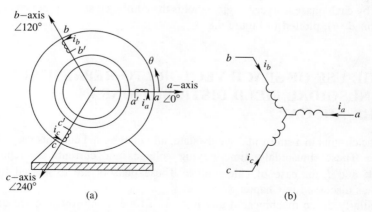

FIGURE 9.7 Three-phase windings.

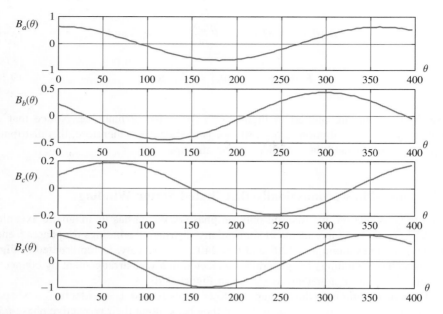

FIGURE 9.8   Waveforms of flux density.

Sample   From Equation 9.8, the peak flux density produced by any phase current i is

$$\hat{B} = \frac{\mu_o N_s}{2\ell_g} i = \frac{4\pi \times 10^{-7} \times 100}{2 \times 1 \times 10^{-3}} i = 0.0628\, i \;\; [\text{T}]$$

The flux-density distributions are plotted as functions of θ in Figure 9.8 for the given values of the three phase currents.

Note that B_a has its positive peak at $\theta = 0°$, B_b has its negative peak at $\theta = 120°$, and B_c has its negative peak at $\theta = 240°$. Applying the principle of superposition under the assumption of a linear magnetic circuit, adding together the flux-density distributions produced by each phase at every angle θ yields the combined stator-produced flux density distribution $B_s(\theta)$, plotted in Figure 9.8.

9.3   THE USE OF SPACE VECTORS TO REPRESENT SINUSOIDAL FIELD DISTRIBUTIONS IN THE AIR GAP

In linear ac circuits in a sinusoidal steady state, all voltages and currents vary sinusoidally with time. These sinusoidally time-varying voltages and currents are represented by phasors $\overline{V}$ and $\overline{I}$ for ease of calculations. These phasors are expressed by complex numbers, as discussed in Chapter 3.

Similarly, in ac machines, at any instant of time t, sinusoidal space distributions of fields (B, H, F) in the air gap can be represented by space vectors. At any instant of time t, in representing a field distribution in the air gap with a space vector, we should note the following:

- The peak of the field distribution is represented by the amplitude of the space vector.
- Where the field distribution has its *positive* peak, the angle θ, measured with respect to the phase-*a* magnetic axis (by convention chosen as the reference axis), is represented by the orientation of the space vector.

Similar to phasors, space vectors are expressed by complex numbers. The space vectors are denoted by a "$\rightarrow$" on top, and their time dependence is explicitly shown.

Let us first consider phase-*a*. In Figure 9.9a, at any instant of time t, the mmf produced by the sinusoidally-distributed phase-*a* winding has a co-sinusoidal shape (distribution) in space; that is, this distribution always peaks along the phase-*a* magnetic axis, and elsewhere it varies with the cosine of the angle θ away from the magnetic axis.

The amplitude of this co-sinusoidal spatial distribution depends on the phase current i_a, which varies with time. Therefore, as shown in Figure 9.9a, at any time t, the mmf distribution due to i_a can be represented by a space vector $\vec{F}_a(t)$:

$$\vec{F}_a(t) = \frac{N_s}{2} i_a(t) \angle 0° \tag{9.14}$$

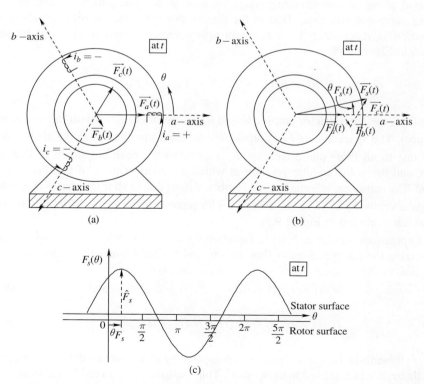

FIGURE 9.9 Representation of MMF space vector in a machine.

The amplitude of $\vec{F}_a(t)$ is $(N_s/2)$ times $i_a(t)$, and $\vec{F}_a(t)$ is always oriented along the phase-a magnetic axis at the angle of $0°$. The phase-a magnetic axis is always used as the reference axis. A representation similar to the mmf distribution can be used for the flux-density distribution.

In a similar manner, at any time t, the mmf distributions produced by the other two phase windings can also be represented by space vectors oriented along their respective magnetic axes at $120°$ and $240°$, as shown in Figure 9.9a for negative values of i_b and i_c. In general, at any instant of time, we have the following three space vectors representing the respective mmf distributions:

$$\vec{F}_a(t) = \frac{N_s}{2} i_a(t) \angle 0°$$

$$\vec{F}_b(t) = \frac{N_s}{2} i_b(t) \angle 120° \qquad (9.15)$$

$$\vec{F}_c(t) = \frac{N_s}{2} i_c(t) \angle 240°$$

Note that the sinusoidal distribution of mmf in the air gap at any time t is a consequence of the sinusoidally distributed windings. As shown in Figure 9.9a for a positive value of i_a and negative values of i_b and i_c (such that $i_a + i_b + i_c = 0$), each of these vectors is pointed along its corresponding magnetic axis, with its amplitude depending on the winding current at that time. Due to the three stator currents, the resultant stator mmf distribution is represented by a resultant space vector, which is obtained by vector addition in Figure 9.9b:

$$\vec{F}_s(t) = \vec{F}_a(t) + \vec{F}_b(t) + \vec{F}_c(t) = \hat{F}_s \angle \theta_{F_s} \qquad (9.16a)$$

where $\hat{F}_s$ is the space vector amplitude and θ_{F_s} is the orientation (with the a-axis as the reference). The space vector $\vec{F}_s(t)$ represents the mmf distribution in the air gap at this time t due to all three phase currents; $\hat{F}_s$ represents the peak amplitude of this distribution, and θ_{F_s} is the angular position at which the positive peak of the distribution is located. The subscript "s" refers to the combined mmf due to all three phases of the stator. The space vector $\vec{F}_s$ at this time in Figure 9.9b represents the mmf distribution in the air gap, which is plotted in Figure 9.9c.

Expressions similar to $\vec{F}_s(t)$ in Equation 9.16a can be derived for the space vectors representing the combined-stator flux-density and the field-intensity distributions:

$$\vec{B}_s(t) = \vec{B}_a(t) + \vec{B}_b(t) + \vec{B}_c(t) = \hat{B}_s \angle \theta_{B_s} \quad \text{and} \qquad (9.16b)$$

$$\vec{H}_s(t) = \vec{H}_a(t) + \vec{H}_b(t) + \vec{H}_c(t) = \hat{H}_s \angle \theta_{H_s} \qquad (9.16c)$$

How are these three field distributions, represented by space vectors defined in Equations 9.16a through 9.16c, related to each other? This question is answered by Equations 9.21a and 9.21b in section 9.4-1.

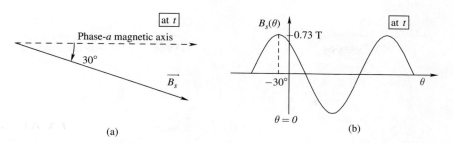

FIGURE 9.10 (a) Resultant flux-density space vector; (b) Flux-density distribution.

Example 9.4

In a 2-pole, three-phase machine, each of the sinusoidally-distributed windings has $N_s = 100$ turns. The air gap length $\ell_g = 1.5$ mm. At a time t, $i_a = 10$ A, $i_b = -10$ A, and $i_c = 0$ A. Using space vectors, calculate and plot the resultant flux density distribution in the air gap at this time.

Sample From Equations 9.15 and 9.16, noting that mathematically $1 \angle 0° = \cos \theta + j \sin \theta$,

$$\vec{F}_s(t) = \frac{N_s}{2} \left(i_a \angle 0° + i_b \angle 120° + i_c \angle 240° \right)$$

$$= 50 \times \left\{ 10 + (-10)(\cos 120° + j \sin 120°) + (0)(\cos 240° + j \sin 240°) \right\}$$

$$= 50 \times 17.32 \angle -30° = 866 \angle -30° \text{ A} \cdot \text{turns}.$$

From Equations 9.8 and 9.9, $B_a(\theta) = (\mu_0/\ell_g)F_a(\theta)$. The same relationship applies to the field quantities due to all three stator phase currents being applied simultaneously; that is, $B_s(\theta) = (\mu_0/\ell_g)F_s(\theta)$. Therefore, at any instant of time t,

$$\vec{B}_s(t) = \frac{\mu_0}{\ell_g} \vec{F}_s(t) = \frac{4\pi \times 10^{-7}}{1.5 \times 10^{-3}} 866 \angle -30° = 0.73 \angle -30° \text{ T}$$

This space vector is drawn in Figure 9.10a. The flux density distribution has a peak value of 0.73 T, and the positive peak is located at $\theta = -30°$, as shown in Figure 9.10b. Elsewhere, the radial flux density in the air gap, due to the combined action of all three phase currents, is co-sinusoidally distributed.

9.4 SPACE-VECTOR REPRESENTATION OF COMBINED TERMINAL CURRENTS AND VOLTAGES

At any time t, we can measure the phase quantities, such as the voltage $v_a(t)$ and the current $i_a(t)$, at the terminals. Since there is no easy way to show that phase currents and voltages are distributed in space at any given time, we will NOT assign space vectors to physically represent these phase quantities. Rather, at any instant of time t, we will define space vectors to mathematically represent the combination of phase voltages and phase currents. These space vectors are defined to be the sum of their phase components (at that time) multiplied by their respective phase-axis orientations. Therefore, at any instant of

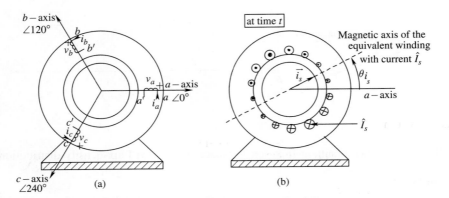

FIGURE 9.11 (a) Phase voltages and currents; (b) Physical interpretation of the current space vector.

time t, the stator current and the stator voltage space vectors are defined, in terms of their phase components (shown in Figure 9.11a), as

$$\vec{i}_s(t) = i_a(t) \angle 0° + i_b(t) \angle 120° + i_c(t) \angle 240° = \hat{I}_s(t) \angle \theta_{i_s}(t) \qquad \text{and} \qquad (9.17)$$

$$\vec{v}_s(t) = v_a(t) \angle 0° + v_b(t) \angle 120° + v_c(t) \angle 240° = \hat{V}_s(t) \angle \theta_{v_s}(t) \qquad (9.18)$$

where the subscript "s" refers to the combined quantities of the stator. We will see later on that this mathematical description is of immense help in understanding the operation and control of ac machines.

9.4.1 Physical Interpretation of the Stator Current Space Vector $\vec{i}_s(t)$

The stator current space vector $\vec{i}_s(t)$ can be easily related to the stator mmf space vector $\vec{F}_s$. Multiplying both sides of Equation 9.17 by $(N_s/2)$ gives

$$\frac{N_s}{2}\vec{i}_s(t) = \underbrace{\frac{N_s}{2}i_a(t) \angle 0°}_{\vec{F}_a(t)} + \underbrace{\frac{N_s}{2}i_b(t) \angle 120°}_{\vec{F}_b(t)} + \underbrace{\frac{N_s}{2}i_c(t) \angle 240°}_{\vec{F}_c(t)} \qquad (9.19a)$$

Using Equation 9.16, the sum of the mmf space vectors for the three phases is the resultant stator space vector. Therefore,

$$\frac{N_s}{2}\vec{i}_s(t) = \vec{F}_s(t) \qquad (9.19b)$$

Thus,

$$\vec{i}_s(t) = \frac{\vec{F}_s(t)}{(N_s/2)} \qquad \text{where} \qquad \hat{I}_s(t) = \frac{\hat{F}_s(t)}{(N_s/2)} \quad \text{and} \quad \theta_{i_s}(t) = \theta_{F_s}(t) \qquad (9.20)$$

Equation 9.20 shows that the vectors $\vec{i}_s(t)$ and $\vec{F}_s(t)$ are related only by a scalar constant $(N_s/2)$. Therefore, they have the same orientation and their amplitudes are related by $(N_s/2)$. At any instant of time t, Equation 9.20 has the following interpretation:

The combined mmf distribution in the air gap produced by i_a, i_b, and i_c flowing through their respective sinusoidally-distributed phase windings (each with N_s turns) is the same as that produced in Figure 9.11b by a current $\hat{I}_s$ flowing through an equivalent sinusoidally-distributed stator winding with its axis oriented at $\theta_{i_s}(t)$. This equivalent winding also has N_s turns.

As we will see later on, the above interpretation is very useful—it allows us to obtain, at any instant of time, the combined torque acting on all three phase windings by calculating the torque acting on this single equivalent winding with a current $\hat{I}_s$.

Next, we will use $\vec{i}_s(t)$ to relate the field quantities produced due to the combined effects of the three stator phase winding currents. Equations 9.7 through 9.9 show that the field distributions H_a, B_a, and F_a, produced by i_a flowing through the phase-a winding, are related by scalar constants. This will also be true for the combined fields in the air gap caused by the simultaneous flow of i_a, i_b, and i_c, since the magnetic circuit is assumed to be unsaturated and the principle of superposition applies. Therefore, we can write expressions for $\vec{B}_s(t)$ and $\vec{H}_s(t)$ in terms of $\vec{i}_s(t)$ which are similar to Equation 9.19b for $\vec{F}_s(t)$ (which is repeated below),

$$\vec{F}_s(t) = \frac{N_s}{2}\vec{i}_s(t)$$

$$\vec{H}_s(t) = \frac{N_s}{2\ell_g}\vec{i}_s(t) \qquad \text{(rotor-circuit electrically open-circuited)} \qquad (9.21\text{a})$$

$$\vec{B}_s(t) = \frac{\mu_o N_s}{2\ell_g}\vec{i}_s(t)$$

The relationships in Equation 9.21a show that these stator space vectors (with the rotor circuit electrically open-circuited) are collinear (that is, they point in the same direction) at any instant of time. Equation 9.21 also yields the relationship between the peak values as

$$\hat{F}_s = \frac{N_s}{2}\hat{I}_s, \hat{H}_s = \frac{N_s}{2\ell_g}\hat{I}_s, \hat{B}_s = \frac{\mu_o N_s}{2\ell_g}\hat{I}_s \qquad \text{(rotor-circuit electrically open-circuited)}$$

$$(9.21\text{b})$$

Example 9.5

For the conditions in an ac machine in Example 9.4 at a given time t, calculate $\vec{i}_s(t)$. Show the equivalent winding and the current necessary to produce the same mmf distribution as the three phase windings combined.

Sample In Example 9.4, $i_a = 10\,\text{A}$, $i_b = -10\,\text{A}$, and $i_c = 0\,\text{A}$. Therefore, from Equation 9.17,

$$\vec{i}_s = i_a \angle 0° + i_b \angle 120° + i_c \angle 240° = 10 + (-10)\angle 120° + (0)\angle 240°$$

$$= 17.32 \angle -30°\,\text{A}$$

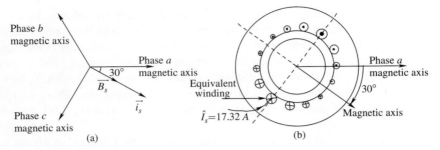

FIGURE 9.12   (a) Stator current space vector; (b) The equivalent winding.

The space vector $\vec{i}_s$ is shown in Figure 9.12a. Since the $\vec{i}_s$ vector is oriented at $\theta = -30°$ with respect to the phase-a magnetic axis, the equivalent sinusoidally-distributed stator winding has its magnetic axis at an angle of $-30°$ with respect to the phase-a winding, as shown in Figure 9.12b. The current required in the equivalent stator winding to produce the equivalent mmf distribution is the peak current $\hat{I}_s = 17.32$ A.

9.4.2   Phase Components of Space Vectors $\vec{i}_s(t)$ and $\vec{v}_s(t)$

If the three stator windings in Figure 9.13a are connected in a wye arrangement, the sum of their currents is zero at any instant of time t by Kirchhoff's Current Law: $i_a(t) + i_b(t) + i_c(t) = 0$.

Therefore, as shown in Figure 9.13b, at any time t, a space vector is constructed from a unique set of phase components, which can be obtained by multiplying the projection of the space vector along the three axes by 2/3. (We should note that if the phase currents were not required to add up to zero, there would be an infinite number of phase component combinations.)

This graphical procedure is based on the mathematical derivations described below. First, let us consider the relationship

$$1 \angle \theta = e^{j\theta} = \cos\theta + j\sin\theta \tag{9.22}$$

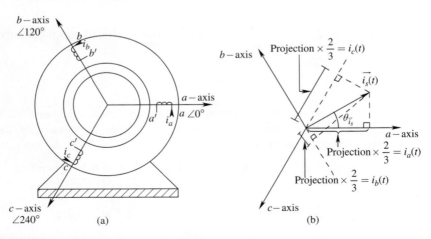

FIGURE 9.13   Phase components of a space vector.

The real part in the above equation is

$$\mathrm{Re}(1 \angle \theta) = \cos \theta \qquad (9.23)$$

Therefore, mathematically, we can obtain the phase components of a space vector such as $\vec{i}_s(t)$ as follows: multiply both sides of the $\vec{i}_s(t)$ expression in Equation 9.17 by $1 \angle 0°$, $1 \angle -120°$, and $1 \angle -240°$, respectively. Equate the real parts on both sides and use the condition that $i_a(t) + i_b(t) + i_c(t) = 0$.

To obtain i_a: $\mathrm{Re}\left[\vec{i}_s \angle 0°\right] = i_a + \underbrace{\mathrm{Re}[i_b \angle 120°]}_{-\frac{1}{2}i_b} + \underbrace{\mathrm{Re}[i_c \angle 240°]}_{-\frac{1}{2}i_c} = \frac{3}{2} i_a$

$$\therefore i_a = \frac{2}{3}\mathrm{Re}\left[\vec{i}_s \angle 0°\right] = \frac{2}{3}\mathrm{Re}\left[\hat{I}_s \angle \theta_{i_s}\right] = \frac{2}{3}\hat{I}_s \cos \theta_{i_s} \qquad (9.24a)$$

To obtain i_b: $\mathrm{Re}\left[\vec{i}_s \angle -120°\right] = \underbrace{\mathrm{Re}[i_a \angle -120°]}_{-\frac{1}{2}i_a} + i_b + \underbrace{\mathrm{Re}[i_c \angle 120°]}_{-\frac{1}{2}i_c} = \frac{3}{2} i_b$

$$\therefore i_b = \frac{2}{3}\mathrm{Re}\left[\vec{i}_s \angle -120°\right] = \frac{2}{3}\mathrm{Re}\left[\hat{I}_s \angle (\theta_{i_s} - 120°)\right] = \frac{2}{3}\hat{I}_s \cos(\theta_{i_s} - 120°) \qquad (9.24b)$$

To obtain i_c: $\mathrm{Re}\left[\vec{i}_s \angle -240°\right] = \underbrace{\mathrm{Re}[i_a \angle -240°]}_{-\frac{1}{2}i_a} + \underbrace{\mathrm{Re}[i_b \angle -120°]}_{-\frac{1}{2}i_b} + i_c = \frac{3}{2} i_c$

$$\therefore i_c = \frac{2}{3}\mathrm{Re}\left[\vec{i}_s \angle -240°\right] = \frac{2}{3}\mathrm{Re}\left[\hat{I}_s \angle (\theta_{i_s} - 240°)\right] = \frac{2}{3}\hat{I}_s \cos(\theta_{i_s} - 240°) \qquad (9.24c)$$

Since $i_a(t) + i_b(t) + i_c(t) = 0$, it can be shown that the same uniqueness applies to components of all space vectors such as $\vec{v}_s(t)$, $\vec{B}_s(t)$, and so on for both the stator and the rotor.

Example 9.6
In an ac machine at a given time, the stator voltage space vector is given as $\vec{v}_s = 254.56 \angle 30°$ V. Calculate the phase voltage components at this time.

Sample From Equation 9.24,

$$v_a = \frac{2}{3}\mathrm{Re}\{\vec{v}_s \angle 0°\} = \frac{2}{3}\mathrm{Re}\{254.56 \angle 30°\} = \frac{2}{3} \times 254.56 \cos 30° = 146.97 \, \mathrm{V},$$

$$v_b = \frac{2}{3}\mathrm{Re}\{\vec{v}_s \angle -120°\} = \frac{2}{3}\mathrm{Re}\{254.56 \angle -90°\} = \frac{2}{3} \times 254.56 \cos(-90°) = 0 \, \mathrm{V}, \text{ and}$$

$$v_c = \frac{2}{3}\mathrm{Re}\{\vec{v}_s \angle -240°\} = \frac{2}{3}\mathrm{Re}\{254.56 \angle -210°\}$$

$$= \frac{2}{3} \times 254.56 \cos(-210°) = -146.97 \, \mathrm{V}.$$

9.5   BALANCED SINUSOIDAL STEADY-STATE EXCITATION (ROTOR OPEN-CIRCUITED)

So far, our discussion has been in very general terms where voltages and currents are not restricted to any specific form. However, we are mainly interested in the normal mode of operation, that is, balanced three-phase, sinusoidal steady state conditions. Therefore, we will assume that a balanced set of sinusoidal voltages at a frequency $f(= \frac{\omega}{2\pi})$ in steady state is applied to the stator, with the rotor assumed to be open-circuited. We will initially neglect the stator winding resistances R_s and the leakage inductances $L_{\ell s}$.

In steady state, applying voltages to the windings in Figure 9.14a (under rotor open-circuit condition) results in magnetizing currents.

These magnetizing currents are indicated by adding "m" to the subscripts in the following equation, and are plotted in Figure 9.14b

$$i_{ma} = \hat{I}_m \cos \omega t, \; i_{mb} = \hat{I}_m \cos (\omega t - 2\pi/3), \text{ and } i_{mc} = \hat{I}_m \cos(\omega t - 4\pi/3) \qquad (9.25)$$

where $\hat{I}_m$ is the peak value of the magnetizing currents and the time origin is chosen to be at the positive peak of $i_{ma}(t)$.

9.5.1   Rotating Stator MMF Space Vector

Substituting into Equation 9.17 the expressions in Equation 9.25 for the magnetizing currents varying sinusoidally with time, the stator magnetizing current space vector is

$$\overrightarrow{i_{ms}}(t) = \hat{I}_m [\cos \omega t \angle 0° + \cos(\omega t - 2\pi/3) \angle 120° + \cos(\omega t - 4\pi/3) \angle 240°] \qquad (9.26)$$

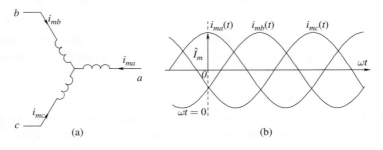

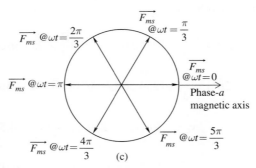

FIGURE 9.14   (a) Windings; (b) Magnetizing currents; (c) Rotating mmf space vector.

The expression within the square bracket in Equation 9.26 simplifies to $\frac{3}{2} \angle \omega t$ (see homework problem 9.8) and Equation 9.26 becomes

$$\vec{i_{ms}}(t) = \frac{3}{2}\hat{I}_m \angle \omega t = \underbrace{\hat{I}_{ms}}_{\hat{I}_{ms}} \angle \omega t \qquad \text{where} \qquad \hat{I}_{ms} = \frac{3}{2}\hat{I}_m \tag{9.27}$$

From Equation 9.21a,

$$\vec{F_{ms}}(t) = \frac{N_s}{2}\vec{i_{ms}}(t) = \hat{F}_{ms} \angle \omega t \quad \text{where} \quad \hat{F}_{ms} = \frac{N_s}{2}\hat{I}_{ms} = \frac{3}{2}\frac{N_s}{2}\hat{I}_m \tag{9.28}$$

Similarly, using Equation 9.21a again,

$$\vec{B_{ms}}(t) = \left(\frac{\mu_o N_s}{2\ell_g}\right)\vec{i_{ms}}(t) \quad \text{where} \quad \hat{B}_{ms} = \left(\frac{\mu_o N_s}{2\ell_g}\right)\hat{I}_{ms} = \frac{3}{2}\left(\frac{\mu_o N_s}{2\ell_g}\right)\hat{I}_m \tag{9.29}$$

Note that if the peak flux density $\hat{B}_{ms}$ in the air gap is to be at its rated value in Equation 9.29, then the $\hat{I}_{ms}$ and hence the peak value of the magnetizing current $\hat{I}_m$ in each phase must also be at their rated values.

Under sinusoidal steady-state conditions, the stator-current, the stator-mmf, and the air gap flux-density space vectors have constant amplitudes ($\hat{I}_{ms}$, $\hat{F}_{ms}$, and $\hat{B}_{ms}$). As shown by $\vec{F_{ms}}(t)$ in Figure 9.14c, all of these space vectors rotate with time at a constant speed, called the synchronous speed ω_{syn}, in the counter-clockwise direction, which in a 2-pole machine is equal to the frequency $\omega\ (= 2\pi f)$ of the voltages and currents applied to the stator:

$$\omega_{syn} = \omega \quad (p = 2) \tag{9.30}$$

Example 9.7
With the rotor electrically open-circuited in a 2-pole ac machine, voltages are applied to the stator, and result in the magnetizing currents plotted in Figure 9.15a. Sketch the direction of the flux lines at the instants $\omega t = 0°$, $60°$, $120°$, $180°$, $240°$, and $300°$. Show that one electrical cycle results in the rotation of the flux orientation by one revolution, in accordance with Equation 9.30 for a 2-pole machine.

Sample At $\omega t = 0$, $i_{ma} = \hat{I}_m$ and $i_{mb} = i_{mc} = -(1/2)\hat{I}_m$. The current directions for the three windings are indicated in Figure 9.15b, where the circles for phase-a are shown larger due to twice as much current in them compared to the other two phases. The resulting flux orientation is shown as well. A similar procedure is followed at other instants, as shown in Figures 9.15c through 9.15g. These drawings clearly show that in a 2-pole machine, the electrical excitation through one cycle of the electrical frequency $f(= \omega/2\pi)$ results in the rotation of the flux orientation, and hence of the space vector $\vec{B_{ms}}$, by one revolution in space. Therefore, $\omega_{syn} = \omega$, as expressed in Equation 9.30.

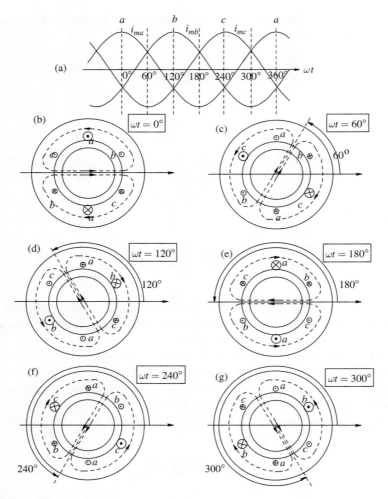

FIGURE 9.15   Example 9.7

9.5.2   Rotating Stator MMF Space Vector in Multi-Pole Machines

In the previous section, we considered a 2-pole machine. In general, in a p-pole machine, a balanced sinusoidal steady state, with currents and voltages at a frequency $f(=\omega/2\pi)$, results in an mmf space vector that rotates at a speed

$$\omega_{syn} = \frac{\omega}{p/2} \qquad \left(\frac{p}{2} = \text{pole} - \text{pairs}\right) \qquad (9.31)$$

This can be illustrated by considering a p-pole machine and repeating the procedure outlined in Example 9.7 for a 2-pole machine (this is left as homework problem 9.11).

In the space vectors for multi-pole machines, the three magnetic axes can be drawn as in a 2-pole machine (similar to the space vector diagrams of Figure 9.9b or 9.13b, for example), except now the axes are separated by 120 degrees (electrical), where the electrical angles are defined by Equation 9.10. Therefore, one complete cycle of electrical excitation causes the space vector, at the synchronous speed given in Equation 9.31, to rotate by 360 degrees (electrical); that is, in the space vector diagram, the space vector returns to the position that it started from. This corresponds to a rotation by an angle of $360/(p/2)$ mechanical degrees, which is exactly what happens within the machine. However, in general (special situations will be pointed out), since no additional insight is gained by this multi-pole representation, it is best to analyze a multi-pole machine as if it were a 2-pole machine.

9.5.3   The Relationship between Space Vectors and Phasors in Balanced Three-Phase Sinusoidal Steady State ($\vec{v_s}\big|_{t=0} \Leftrightarrow \overline{V}_a$ and $\vec{i_{ms}}\big|_{t=0} \Leftrightarrow \overline{I}_{ma}$)

In Figure 9.14b, note that at $\omega t = 0$, the magnetizing current i_{ma} in phase-a is at its positive peak. Corresponding to this time $\omega t = 0$, the space vectors $\vec{i_{ms}}$, $\vec{F_{ms}}$, and $\vec{B_{ms}}$ are along the a-axis in Figure 9.14c. Similarly, at $\omega t = 2\pi/3\ rad$ or $120°$, i_{mb} in phase-b reaches its positive peak. Correspondingly, the space vectors $\vec{i_{ms}}$, $\vec{F_{ms}}$, and $\vec{B_{ms}}$ are along the b-axis, $120°$ ahead of the a-axis. Therefore, we can conclude that under a balanced three-phase sinusoidal steady state, when a phase voltage (or a phase current) is at its positive peak, the combined stator voltage (or current) space vector will be oriented along that phase axis. This can also be stated as follows: when a combined stator voltage (or current) space vector is oriented along the magnetic axis of any phase, at that time, that phase voltage (or current) is at its positive peak value.

We will make use of the information in the above paragraph. Under a balanced three-phase sinusoidal steady state, let us arbitrarily choose some time as the origin $t = 0$ in Figure 9.16a such that the current i_{ma} reaches its positive peak at a later time $\omega t = \alpha$. The phase-a current can be expressed as

$$i_{ma}(t) = \hat{I}_m \cos{(\omega t - \alpha)} \tag{9.32}$$

which is represented by a phasor below and shown in the phasor diagram of Figure 9.16b:

$$\overline{I}_{ma} = \hat{I}_m \angle -\alpha \tag{9.33a}$$

The phase-a current $i_{ma}(t)$ reaches its positive peak at $\omega t = \alpha$. Therefore, at time $t = 0$, the $\vec{i_{ms}}$ space vector will be as shown in Figure 9.16c, behind the magnetic axis of phase-a by an angle α, so that it will be along the a-axis at a later time $\omega t = \alpha$, when i_{ma} reaches its positive peak. Therefore, at time $t = 0$,

$$\vec{i_{ms}}\big|_{t=0} = \hat{I}_{ms} \angle -\alpha \quad \text{where} \quad \hat{I}_{ms} = \frac{3}{2}\hat{I}_m \tag{9.33b}$$

Combining Equations 9.33a and 9.33b,

$$\vec{i_{ms}}\big|_{t=0} = \frac{3}{2}\overline{I}_{ma} \tag{9.34}$$

where the left side mathematically represents the combined current space vector at time $t = 0$ and in the right side $\overline{I}_{ma}$ is the phase-a current phasor representation. In sinusoidal

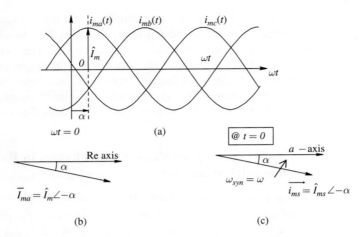

FIGURE 9.16 Relationship between space vectors and phasors in balanced sinusoidal steady state.

steady state, Equation 9.34 illustrates an important relationship between space vectors and phasors which we will use very often:

1. The orientation of the phase-*a* voltage (or current) phasor is the same as the orientation of the combined stator voltage (or current) space vector at time $t = 0$.
2. The amplitude of the combined stator voltage (or current) space vector is larger than that of the phasor amplitude by a factor of $3/2$.

Note that knowing the phasors for phase-*a* is sufficient, as the other phase quantities are displaced by 120 degrees with respect to each other and have equal magnitudes. This concept will be used in the following section.

9.5.4 Induced Voltages in Stator Windings

In the following discussion, we will ignore the resistance and the leakage inductance of the stator windings, shown wye-connected in Figure 9.17a. Neglecting all losses, under the condition that there is no electrical circuit or excitation in the rotor, the stator windings appear purely inductive. Therefore, in each phase, the phase voltage and the magnetizing current are related as

$$e_{ma} = L_m \frac{di_{ma}}{dt}, \quad e_{mb} = L_m \frac{di_{mb}}{dt}, \quad \text{and} \quad e_{mc} = L_m \frac{di_{mc}}{dt} \tag{9.35}$$

where L_m is the magnetizing inductance of the three-phase stator, which in terms of the machine parameters can be calculated as (see homework problems 9.13 and 9.14)

$$L_m = \frac{3}{2} \left[\frac{\pi \mu_o r \ell}{\ell_g} \left(\frac{N_s}{2} \right)^2 \right] \tag{9.36}$$

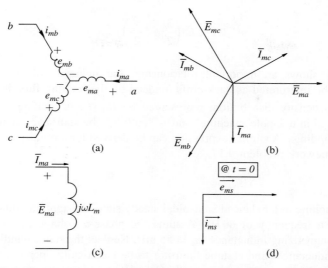

FIGURE 9.17 Winding current and induced emf (a) Individual windings; (b) Phasors; (c) Per-phase equivalent circuit (d) Space vectors.

where r is the radius, ℓ is the rotor length, and ℓ_g is the air gap length. The combination of quantities within the square bracket is the single-phase self-inductance $L_{m,1-phase}$ of each of the stator phase windings in a 2-pole machine:

$$L_{m,1-phase} = \frac{\pi\mu_o r\ell}{\ell_g}\left(\frac{N_s}{2}\right)^2 \tag{9.37}$$

Due to mutual coupling between the three phases, L_m given in Equation 9.36 is larger than $L_{m,1-phase}$ by a factor of 3/2:

$$L_m = \frac{3}{2}L_{m,1-phase} \tag{9.38}$$

Under a balanced sinusoidal steady state, assuming that i_{ma} peaks at $\omega t = 90°$, we can draw the three-phase phasor diagram shown in Figure 9.17b, where

$$\overline{E}_{ma} = (j\omega L_m)\overline{I}_{ma} \tag{9.39}$$

The phasor-domain circuit diagram for phase-a is shown in Figure 9.17c, and the corresponding combined space vector diagram for $\vec{e}_{ms}$ and $\vec{i}_{ms}$ at $t = 0$ is shown in Figure 9.17d. In general, at any time t,

$$\vec{e}_{ms}(t) = (j\omega L_m)\vec{i}_{ms}(t) \quad \text{where} \quad \hat{E}_{ms} = (\omega L_m)\hat{I}_{ms} = \frac{3}{2}(\omega L_m)\hat{I}_m \tag{9.40}$$

In Equation 9.40, substituting for $\vec{i}_{ms}(t)$ in terms of $\vec{B}_{ms}(t)$ from Equation 9.21a and substituting for L_m from Equation 9.36,

$$\vec{e}_{ms}(t) = j\omega \left(\frac{3}{2}\pi r \ell \frac{N_s}{2}\right) \vec{B}_{ms}(t) \tag{9.41}$$

Equation 9.41 shows an important relationship: the induced voltages in the stator windings can be interpreted as back-emfs induced by the rotating flux-density distribution. This flux-density distribution, represented by $\vec{B}_{ms}(t)$, is rotating at a speed ω_{syn} (which equals ω in a 2-pole machine) and is "cutting" the stationary conductors of the stator phase windings. A similar expression can be derived for a multi-pole machine with $p > 2$ (see homework problem 9.17).

Example 9.8

In a 2-pole machine in a balanced sinusoidal steady state, the applied voltages are 208 V (L-L, rms) at a frequency of 60 Hz. Assume the phase-*a* voltage to be the reference phasor. The magnetizing inductance $L_m = 55$ mH. Neglect the stator winding resistances and leakage inductances and assume the rotor to be electrically open-circuited. (a) Calculate and draw the $\overline{E}_{ma}$ and $\overline{I}_{ma}$ phasors. (b) Calculate and draw the space vectors $\vec{e}_{ms}$ and $\vec{i}_{ms}$ at $\omega t = 0°$ and $\omega t = 60°$. (c) If the peak flux density in the air gap is 1.1 T, draw the $\vec{B}_{ms}$ space vector in part (b) at the two instants of time.

Sample

(a) With the phase-*a* voltage as the reference phasor,

$$\overline{E}_{ma} = \frac{208\sqrt{2}}{\sqrt{3}} \angle 0° = 169.83 \angle 0° \text{ V and}$$

$$\overline{I}_{ma} = \frac{\overline{E}_{ma}}{j\omega L_m} \angle 0° = \frac{169.83}{2\pi \times 60 \times 55 \times 10^{-3}} \angle -90° = 8.19 \angle -90° \text{ A}$$

These two phasors are drawn in Figure 9.18a.

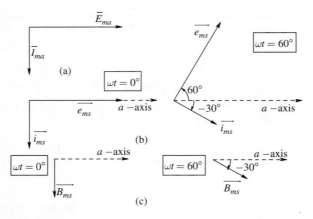

FIGURE 9.18   Example 9.7

(b) At $\omega t = 0°$, from Equation 9.34, as shown in Figure 9.18b,

$$\vec{e_{ms}}\Big|_{\omega t=o} = \frac{3}{2}\overline{E}_{ma} = \frac{3}{2}169.83 \angle 0° = 254.74 \angle 0° \text{ V and}$$

$$\vec{i_{ms}}\Big|_{\omega t=0} = \frac{3}{2}\overline{I}_{ma} = \frac{3}{2}8.19 \angle -90° = 12.28 \angle -90° \text{ A}$$

At $\omega t = 60°$, both space vectors have rotated by an angle of 60 degrees in a counter-clockwise direction, as shown in Figure 9.18b. Therefore,

$$\vec{e_{ms}}\Big|_{\omega t=60°} = \vec{e_{ms}}\Big|_{\omega t=0}(1 \angle 60°) = 254.74 \angle 60° \text{ V and}$$

$$\vec{i_{ms}}\Big|_{\omega t=60°} = \vec{i_{ms}}\Big|_{\omega t=0}(1 \angle 60°) = 12.28 \angle -30° \text{ A}$$

(c) In this example, at any time, the stator flux density space vector $\vec{B_{ms}}$ is oriented in the same direction as the $\vec{i_{ms}}$ space vector. Therefore, as plotted in Figure 9.18c,

$$\vec{B_{ms}}\Big|_{\omega t=0°} = 1.1 \angle -90° \text{ T} \quad \text{and} \quad \vec{B_{ms}}\Big|_{\omega t=60°} = 1.1 \angle -30° \text{ T}$$

SUMMARY/REVIEW QUESTIONS

1. Draw the three-phase axis in the motor cross-section. Also, draw the three phasors $\overline{V}_a$, $\overline{V}_b$, and $\overline{V}_c$ in a balanced sinusoidal steady state. Why is the phase-b axis ahead of the phase-a axis by 120 degrees, but $\overline{V}_b$ lags $\overline{V}_a$ by 120 degrees?
2. Ideally, what should be the field (F, H, and B) distributions produced by each of the three stator windings? What is the direction of this field in the air gap? What direction is considered positive, and what is considered negative?
3. What should the conductor-density distribution in a winding be in order to achieve the desired field distribution in the air gap? Express the conductor-density distribution $n_s(\theta)$ for phase-a.
4. How is sinusoidal distribution of conductor density in a phase winding approximated in practical machines with only a few slots available to each phase?
5. How are the three field distributions (F, H, and B) related to each other, assuming that there is no magnetic saturation in the stator and the rotor iron?
6. What is the significance of the magnetic axis of any phase winding?
7. Mathematically express the field distributions in the air gap due to i_a as a function of θ. Repeat this for i_b and i_c.
8. What do the phasors $\overline{V}$ and $\overline{I}$ denote? What are the meanings of the space vectors $\vec{B}_a(t)$ and $\vec{B}_s(t)$ at time t, assuming that the rotor circuit is electrically open-circuited?
9. What is the constraint on the sum of the stator currents?
10. What are physical interpretations of various stator winding inductances?
11. Why is the per-phase inductance L_m greater than the single-phase inductance $L_{m,1-phase}$ by a factor of 3/2?
12. What are the characteristics of space vectors that represent the field distributions $F_s(\theta)$, $H_s(\theta)$, and $B_s(\theta)$ at a given time? What notations are used for these space vectors? Which axis is used as a reference to express them mathematically in this chapter?

13. Why does a dc current through a phase winding produce a sinusoidal flux-density distribution in the air gap?
14. How are the terminal phase voltages and currents combined for representation by space vectors?
15. What is the physical interpretation of the stator current space vector $\vec{i}_s(t)$?
16. With no excitation or currents in the rotor, are all of the space vectors associated with the stator $\vec{i}_{ms}(t), \vec{F}_{ms}(t), \vec{B}_{ms}(t)$ collinear (oriented in the same direction)?
17. In ac machines, a stator space vector $\vec{v}_s(t)$ or $\vec{i}_s(t)$ consists of a unique set of phase components. What is the condition on which these components are based?
18. Express the phase voltage components in terms of the stator voltage space vector.
19. Under three-phase balanced sinusoidal condition with no rotor currents, and neglecting the stator winding resistances R_s and the leakage inductance $L_{\ell s}$ for simplification, answer the following questions: (a) What is the speed at which all of the space vectors rotate? (b) How is the peak flux density related to the magnetizing currents? Does this relationship depend on the frequency f of the excitation? If the peak flux density is at its rated value, then what about the peak value of the magnetizing currents? (c) How do the magnitudes of the applied voltages depend on the frequency of excitation, in order to keep the flux density constant (at its rated value for example)?
20. What is the relationship between space vectors and phasors under balanced sinusoidal operating conditions?

REFERENCES

1. A.E. Fitzgerald, Charles Kingsley, and Umans, *Electric Machinery*, 5th ed. (New York: McGraw Hill, 1990).
2. G. R. Slemon, *Electric Machines and Drives* (Addison-Wesley, Inc., 1992).
3. P. K. Kovacs, *Transient Phenomena in Electrical Machines* (Elsevier, 1984).

PROBLEMS

9.1   In a three-phase, 2-pole ac machine, assume that the neutral of the wye-connected stator windings is accessible. The rotor is electrically open-circuited. The phase-a is applied a current $i_a(t) = 10 \sin \omega t$. Calculate $\vec{B}_a$ at the following instants of ωt: 0, 90, 135, and 210 degrees. Also, plot the $B_a(\theta)$ distribution at these instants.

9.2   In the sinusoidal conductor-density distribution shown in Figure 9.3, make use of the symmetry at θ and at $(\pi - \theta)$ to calculate the field distribution $H_a(\theta)$ in the air gap.

9.3   In ac machines, why is the stator winding for phase-b placed 120 degrees ahead of phase-a (as shown in Figure 9.1), whereas the phasors for phase-b (such as $\overline{V}_b$) lag behind the corresponding phasors for phase-a?

9.4   In Example 9.2, derive the expression for $n_s(\theta_e)$ for a 4-pole machine. Generalize it for a multi-pole machine.

9.5   In Example 9.2, obtain the expressions for $H_a(\theta_e)$, $B_a(\theta_e)$, and $F_a(\theta_e)$.

9.6   In a 2-pole, three-phase machine with $N_s = 100$, calculate $\vec{i}_s$ and $\vec{F}_s$ at a time t if at that time the stator currents are as follows: (a) $i_a = 10\,\text{A}$, $i_b = -5\,\text{A}$, and $i_c = -5\,\text{A}$; (b) $i_a = -5\,\text{A}$, $i_b = 10\,\text{A}$, and $i_c = -5\,\text{A}$; (c) $i_a = -5\,\text{A}$, $i_b = -5\,\text{A}$, and $i_c = 10\,\text{A}$.

9.7 In a wye-connected stator, at a time t, $\vec{v}_s = 150 \angle -30°$ V. Calculate v_a, v_b, and v_c at that time.

9.8 Show that the expression in the square brackets of Equation 9.26 simplifies to $\frac{3}{2} \angle \omega t$.

9.9 In a 2-pole, three-phase ac machine, $\ell_g = 1.5$ mm and $N_s = 100$. During a balanced, sinusoidal, 60-Hz steady state with the rotor electrically open-circuited, the peak of the magnetizing current in each phase is 10 A. Assume that at $t = 0$, the phase-a current is at its positive peak. Calculate the flux-density distribution space vector as a function of time. What is the speed of its rotation?

9.10 In Problem 9.9, what would be the speed of rotation if the machine had 6 poles?

9.11 By means of drawings similar to Example 9.7, show the rotation of the flux lines, and hence the speed, in a 4-pole machine.

9.12 In a three-phase ac machine, $\overline{V}_a = 120\sqrt{2} \angle 0°$ V and $\overline{I}_{ma} = 5\sqrt{2} \angle -90°$ A. Calculate and draw $\vec{e}_{ms}$ and $\vec{i}_{ms}$ space vectors at $t = 0$. Assume a balanced, sinusoidal, three-phase steady-state operation at 60 Hz. Neglect the resistance and the leakage inductance of the stator phase windings.

9.13 Show that in a 2-pole machine, $L_{m,1-phase} = \frac{\pi \mu_o r \ell}{\ell_g} \left(\frac{N_s}{2} \right)^2$.

9.14 Show that $L_m = \frac{3}{2} L_{m,1-phase}$.

9.15 In a three-phase ac machine, $\overline{V}_a = 120\sqrt{2} \angle 0°$ V. The magnetizing inductance $L_m = 75$ mH. Calculate and draw the three magnetizing current phasors. Assume a balanced, sinusoidal, three-phase steady-state operation at 60 Hz.

9.16 In a 2-pole, three-phase ac machine, $\ell_g = 1.5$ mm, $\ell = 24$ cm, $r = 6$ cm, and $N_s = 100$. Under a balanced, sinusoidal, 60-Hz steady state, the peak of the magnetizing current in each phase is 10 A. Assume that at $t = 0$, the current in phase-a is at its positive peak. Calculate the expressions for the induced back-emfs in the three-stator phases.

9.17 Recalculate Equation 9.41 for a multi-pole machine with $p > 2$.

9.18 Calculate L_m in a p-pole machine ($p \geq 2$).

9.19 Combine the results of Problems 9.17 and 9.18 to show that for $p \geq 2$, $\vec{e}_{ms}(t) = j\omega L_m \vec{i}_{ms}(t)$.

9.20 In Figure 9.13 of Example 9.7, plot the flux-density distributions produced by each of the phases in parts (b) through (g).

9.21 At some instant of time, $\vec{B}_s(t) = 1.1 \angle 30°$ T. Calculate and plot the flux-density distribution produced by each of the phases as a function of θ.

9.22 In Equation 9.41, the expression for $\vec{e}_{ms}(t)$ is obtained by using the expression of inductance in Equation 9.36. Instead of following this procedure, calculate the voltages induced in each of the stator phases due to the rotating $\vec{B}_{ms}$ to confirm the expression of $\vec{e}_{ms}(t)$ in Equation 9.41.

10

SINUSOIDAL PERMANENT MAGNET AC (PMAC) DRIVES, LCI-SYNCHRONOUS MOTOR DRIVES, AND SYNCHRONOUS GENERATORS

10.1 INTRODUCTION

Having been introduced to ac machines and their analysis using space vector theory, we will now study an important class of ac drives, namely sinusoidal-waveform, permanent-magnet ac (PMAC) drives. The motors in these drives have three phase, sinusoidally-distributed ac stator windings, and the rotor has dc excitation in the form of permanent magnets. We will examine these machines for servo applications, usually in small (<10 kW) power ratings. In such drives, the stator windings of the machine are supplied by controlled currents, which require a closed-loop operation, as shown in the block diagram of Figure 10.1.

These drives are also related to the ECM drives of Chapter 7. The difference here is the sinusoidally-distributed nature of the stator windings that are supplied by sinusoidal wave-form currents. Also, the permanent magnets on the rotor are shaped to induce (in the stator windings) back-emfs that are ideally sinusoidally-varying with time. Unlike the ECM drives, PMAC drives are capable of producing a smooth torque, and thus they are

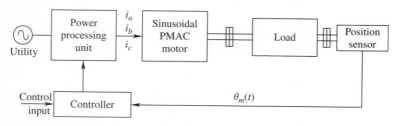

FIGURE 10.1 Block diagram of the closed loop operation of a PMAC drive.

174

used in high-performance applications. They do not suffer from the maintenance problems associated with brush-type dc machines. They are also used where a high efficiency and a high power density are required.

PMAC drives used in low power ratings are in principle similar to synchronous-motor drives used in very large power ratings (in excess of one megawatt) in applications such as controlling the speed of induced-draft fans and boiler feed-water pumps in the central power plants of electric utilities. Such synchronous-motor drives are briefly described in section 10.5.

The discussion of PMAC drives also lends itself to the analysis of line-connected synchronous machines, which are used in very large ratings in the central power plants of utilities to generate electricity. We will briefly analyze these synchronous generators in section 10.6.

10.2 THE BASIC STRUCTURE OF PERMANENT-MAGNET AC (PMAC) MACHINES

We will first consider 2-pole machines, like the one shown schematically in Figure 10.2a, and then we will generalize our analysis to p-pole machines where $p > 2$. The stator contains three-phase, wye-connected, sinusoidally-distributed windings (discussed in Chapter 9), which are shown in the cross-section of Figure 10.2a. These sinusoidally-distributed windings produce a sinusoidally-distributed mmf in the air gap.

10.3 PRINCIPLE OF OPERATION

10.3.1 Rotor-Produced Flux Density Distribution

The permanent-magnet pole pieces mounted on the rotor surface are shaped to ideally produce a sinusoidally-distributed flux density in the air gap. Without delving into detailed construction, Figure 10.2a schematically shows a two-pole rotor. Flux lines leave

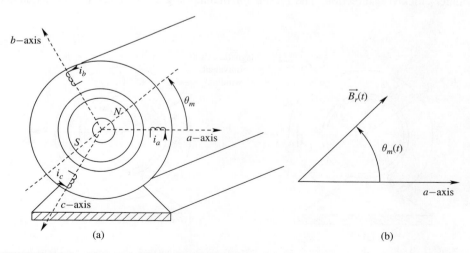

(a)   (b)

FIGURE 10.2 Two-pole PMAC machine.

the rotor at the north pole to re-enter the air gap at the south pole. The rotor-produced flux-density distribution in the air gap (due to flux lines that completely cross the two air gaps) has its positive peak $\hat{B}_r$ directed along the north pole axis. Because this flux density is sinusoidally distributed, it can be represented, as shown in Figure 10.2b, by a space vector of length $\hat{B}_r$, and its orientation can be established by the location of the positive peak of the flux-density distribution. As the rotor turns, the entire rotor-produced flux density distribution in the air gap rotates with it. Therefore, using the stationary stator phase-*a* axis as the reference, we can represent the rotor-produced flux density space vector at a time *t* as

$$\vec{B}_r(t) = \hat{B}_r \angle \theta_m(t) \tag{10.1}$$

where the rotor flux-density distribution axis is at an angle $\theta_m(t)$ with respect to the *a*-axis. In Equation 10.1, permanent magnets produce a constant $\hat{B}_r$, but $\theta_m(t)$ is a function of time, as the rotor turns.

10.3.2 Torque Production

We would like to compute the electromagnetic torque produced by the rotor. However, the rotor consists of permanent magnets, and we have no direct way of computing this torque. Therefore, we will first calculate the torque exerted on the stator; this torque is transferred to the motor foundation. The torque exerted on the rotor is equal in magnitude to the stator torque but acts in the opposite direction.

An important characteristic of the machines under consideration is that they are supplied through the power-processing unit shown in Figure 10.1, which controls the currents $i_a(t)$, $i_b(t)$, and $i_c(t)$ supplied to the stator at any instant of time. At any time *t*, the three stator currents combine to produce a stator current space vector $\vec{i}_s(t)$, which is controlled to be ahead of (or leading) the space vector $\vec{B}_r(t)$ by an angle of 90° in the direction of rotation, as shown in Figure 10.3a. This produces a torque on the rotor in a counter-clockwise direction. The reason for maintaining a 90° angle will be explained

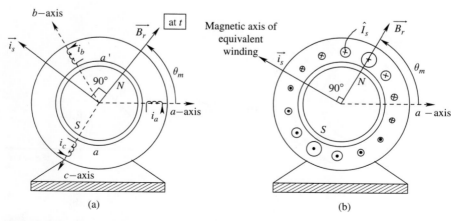

(a)   (b)

FIGURE 10.3   The stator current and the rotor field space vectors in PMAC drives.

shortly. With the *a*-axis as the reference axis, the stator current space vector can be expressed as

$$\vec{i}_s(t) = \hat{I}_s(t) \angle \theta_{i_s}(t) \quad \text{Where} \quad \theta_{i_s}(t) = \theta_m(t) + 90° \qquad (10.2)$$

During a steady-state operation, $\hat{I}_s$ is kept constant while $\theta_m(=\omega_m t)$ changes linearly with time.

We have seen the physical interpretation of the current space vector $\vec{i}_s(t)$ in Chapter 9. In Figure 10.3a at a time t, the three stator phase currents combine to produce an mmf distribution in the air gap.

This mmf distribution is the same as that produced in Figure 10.3b by a single equivalent stator winding that has N_s sinusoidally-distributed turns supplied by a current $\hat{I}_s$ and that has its magnetic axis situated along the $\vec{i}_s(t)$ vector. As seen from Figure 10.3b, by controlling the stator current space vector $\vec{i}_s(t)$ to be 90° ahead of $\vec{B}_r(t)$, all of the conductors in the equivalent stator winding will experience a force acting in the same direction, which in this case is clockwise on the stator (and hence produces a counter-clockwise torque on the rotor). This justifies the choice of 90°: it results in the maximum torque per ampere of stator current because at any other angle some conductors will experience a force in the direction opposite that on other conductors, a condition that will result in a smaller net torque.

As $\vec{B}_r(t)$ rotates with the rotor, the space vector $\vec{i}_s(t)$ is made to rotate at the same speed, maintaining a "lead" of 90°. Thus, the torque developed in the machine of Figure 10.3 depends only on $\hat{B}_r$ and $\hat{I}_s$, and is independent of θ_m. Therefore, to simplify our calculation of this torque in terms of the machine parameters, we will redraw Figure 10.3b as in Figure 10.4 by assuming $\theta_m = 0°$.

Using the expression for force ($f_{em} = B \ell i$), we can calculate the clockwise torque acting on the stator as follows: in the equivalent stator winding shown in Figure 10.4, at an angle ξ, the differential angle $d\xi$ contains $n_s(\xi) \cdot d\xi$ conductors. Using Equation 9.5 and noting that the angle ξ is measured here from the location of the peak conductor density, the conductor density $n_s(\xi) = (N_s/2) \cdot \cos \xi$. Therefore,

The number of conductors in the differential angle $d\xi = \dfrac{N_s}{2} \cos \xi \cdot d\xi$ \qquad (10.3)

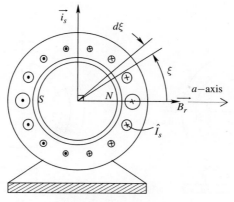

FIGURE 10.4 Torque calculation on the stator.

The rotor-produced flux density at angle ξ is $\hat{B}_r \cos \xi$. Therefore, the torque $dT_{em}(\xi)$ produced by these conductors (due to the current $\hat{I}_s$ flowing through them) located at angle ξ, at a radius r, and of length ℓ is

$$dT_{em}(\xi) = r \cdot \underbrace{\hat{B}_r \cos \xi}_{\text{flux density at } \xi} \cdot \underbrace{\ell}_{\text{cond. length}} \cdot \hat{I}_s \cdot \underbrace{\frac{N_s}{2} \cos \xi \cdot d\xi}_{\text{no. of cond. in } d\xi} \qquad (10.4)$$

To account for the torque produced by all of the stator conductors, we will integrate the above expression from $\xi = -\pi/2$ to $\xi = \pi/2$, and then multiply by a factor of 2, making use of symmetry:

$$T_{em} = 2 \times \int_{\xi=-\pi/2}^{\xi=\pi/2} dT_{em}(\xi) = 2 \frac{N_s}{2} r \ell \hat{B}_r \hat{I}_s \int_{-\pi/2}^{\pi/2} \cos^2 \xi \cdot d\xi = \left(\pi \frac{N_s}{2} r \ell \hat{B}_r \right) \hat{I}_s \quad (10.5)$$

In the above equation, all quantities within the brackets, including $\hat{B}_r$ in a machine with permanent magnets, depend on the machine design parameters and are constants. As noted earlier, the electromagnetic torque produced by the rotor is equal to that in Equation 10.5 in the opposite direction (counter-clockwise in this case). This torque in a 2-pole machine can be expressed as

$$T_{em} = k_T \hat{I}_s \qquad \text{where} \qquad k_T = \pi \frac{N_s}{2} r \ell \hat{B}_r \ (p=2) \qquad (10.6)$$

In the above equation, k_T is the *machine torque constant*, which has the units of Nm/A. Equation 10.6 shows that by controlling the stator phase currents so that the corresponding stator current space vector is ahead (in the desired direction) of the rotor-produced flux-density space vector by 90°, the torque developed is only proportional to $\hat{I}_s$. This torque expression is similar to that in the brush-type dc-motor drives of Chapter 7.

The similarities between the brush-type dc motor drives of Chapter 7 and the PMAC motor drives are shown by means of Figure 10.5.

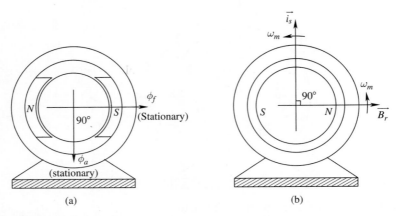

FIGURE 10.5   Similarities between (a) DC motor and (b) PMAC motor drives.

In the brush-type dc motors, the flux ϕ_f produced by the stator and the armature flux ϕ_a produced by the armature winding remain directed orthogonal (at 90°) to each other, as shown in Figure 10.5a. The stator flux ϕ_f is stationary, and so is ϕ_a (due to the commutator action), even though the rotor is turning. The torque produced is controlled by the armature current $i_a(t)$. In PMAC motor drives, the stator-produced flux-density $\vec{B}_{s,\vec{i}_s}(t)$ due to $\vec{i}_s(t)$ is controlled to be directed orthogonal (at 90° in the direction of rotation) to the rotor flux-density $\vec{B}_r(t)$, as shown in Figure 10.5b. Both of these space vectors rotate at the speed ω_m of the rotor, maintaining the 90° angle between the two. The torque is controlled by the magnitude $\hat{I}_s(t)$ of the stator-current space vector.

At this point, we should note that PMAC drives constitute a class that we will call *self-synchronous* motor drives, where the speed of the stator-produced mmf distribution is synchronized to be equal the mechanical speed of the rotor. This feature characterizes a machine as a synchronous machine. The term *self* is added to distinguish these machines from the conventional synchronous machines described in section 10.6. In PMAC drives, this synchronism is established by a closed feedback loop in which the measured instantaneous position of the rotor directs the power-processing unit to locate the stator mmf distribution 90 degrees ahead of the rotor-field distribution. Therefore, there is no possibility of losing synchronism between the two, unlike in the conventional synchronous machines of section 10.6.

10.3.2.1 Generator Mode

PMAC drives can be operated as generators. In fact, these drives are used in this mode in wind-turbines. In this mode, the stator current space vector $\vec{i}_s$ is controlled to be 90 degrees behind the rotor flux-density vector, in the direction of rotation, in Figure 10.3. Therefore, the resulting electromagnetic torque T_{em}, as calculated by Equation 10.5, acts in a direction to oppose the direction of rotation.

10.3.3 Mechanical System of PMAC Drives

The electromagnetic torque acts on the mechanical system connected to the rotor, as shown in Figure 10.6, and the resulting speed ω_m can be obtained from the equation below:

$$\frac{d\omega_m}{dt} = \frac{T_{em} - T_L}{J_{eq}} \tag{10.7}$$

where J_{eq} is the combined motor-load inertia and T_L is the load torque, which may include friction. The rotor position $\theta_m(t)$ is

$$\theta_m(t) = \theta_m(0) + \int_o^t \omega_m(\tau) \cdot d\tau \,(\tau = \text{variable of integration}) \tag{10.8}$$

where $\theta_m(0)$ is the rotor position at time $t = 0$.

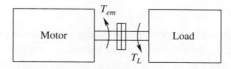

FIGURE 10.6 Rotor-load mechanical system.

10.3.4   Calculation of the Reference Values $i_a^*(t)$, $i_b^*(t)$, and $i_c^*(t)$ of the Stator Currents

The controller in Figure 10.1 is responsible for controlling the torque, speed, and position of the mechanical system. It does so by calculating the instantaneous value of the desired (reference) torque $T_{em}^*(t)$ that the motor must produce. The reference torque may be generated by the cascaded controller discussed in Chapter 8. From Equation 10.6, $\hat{I}_s^*(t)$, the reference value of the amplitude of the stator-current space vector, can be calculated as

$$\hat{I}_s^*(t) = \frac{T_{em}^*(t)}{k_T} \tag{10.9}$$

where k_T is the motor torque constant given in Equation 10.6 (k_T is usually listed in the motor specification sheet).

The controller in Figure 10.1 receives the instantaneous rotor position θ_m, which is measured, as shown in Figure 10.1, by means of a mechanical sensor such as a resolver or an optical encoder (with some restrictions), as discussed in Chapter 17.

With $\theta_m(t)$ as one of the inputs and $\hat{I}_s^*(t)$ calculated from Equation 10.9, the instantaneous reference value of the stator-current space vector becomes

$$\vec{i}_s^{*}(t) = \hat{I}_s^*(t) \angle \theta_{i_s}^*(t) \qquad \text{where} \quad \theta_{i_s}^*(t) = \theta_m(t) + \frac{\pi}{2} \quad \text{(2-pole)} \tag{10.10}$$

Equation 10.10 assumes a 2-pole machine and the desired rotation to be in the counter-clockwise direction. For a clockwise rotation, the angle $\theta_{i_s}^*(t)$ in Equation 10.10 will be $\theta_m(t) - \pi/2$. In a multi-pole machine with $p > 2$, the electrical angle $\theta_{i_s}^*(t)$ will be

$$\theta_{i_s}^*(t) = \frac{p}{2}\theta_m(t) \pm \frac{\pi}{2} \qquad (p \geq 2) \tag{10.11}$$

where $\theta_m(t)$ is the mechanical angle. From $\vec{i}_s^{*}(t)$ in Equation 10.10 (with Equation 10.11 for $\theta_{i_s}^*(t)$ in a machine with $p > 2$), the instantaneous reference values $i_a^*(t)$, $i_b^*(t)$, and $i_c^*(t)$ of the stator phase currents can be calculated using the analysis in the previous chapter (Equations 9.24a through 9.24c):

$$i_a^*(t) = \frac{2}{3}\text{Re}\left[\vec{i}_s^{*}(t)\right] = \frac{2}{3}\hat{I}_s^*(t)\cos\theta_{i_s}^*(t) \tag{10.12a}$$

$$i_b^*(t) = \frac{2}{3}\text{Re}\left[\vec{i}_s^{*}(t)\angle -\frac{2\pi}{3}\right] = \frac{2}{3}\hat{I}_s^*(t)\cos\left(\theta_{i_s}^*(t) - \frac{2\pi}{3}\right) \quad \text{and} \tag{10.12b}$$

$$i_c^*(t) = \frac{2}{3}\text{Re}\left[\vec{i}_s^{*}(t)\angle -\frac{4\pi}{3}\right] = \frac{2}{3}\hat{I}_s^*(t)\cos\left(\theta_{i_s}^*(t) - \frac{4\pi}{3}\right) \tag{10.12c}$$

Section 10.4, which deals with the power-processing unit and the controller, describes how the phase currents, based on the above reference values, are supplied to the motor. Equations 10.12a through 10.12c show that in the balanced sinusoidal steady state, the currents have the

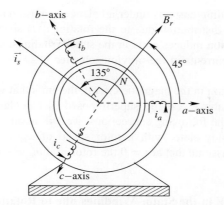

FIGURE 10.7   Stator current space vector for Example 10-1.

constant amplitude of $\hat{I}_s^*$; they vary sinusoidally with time as the angle $\theta_{i_s}^*(t)$ in Equation 10.10 or Equation 10.11 changes continuously with time at a constant speed ω_m:

$$\theta_{i_s}^*(t) = \frac{p}{2}[\theta_m(0) + \omega_m t] \pm \frac{\pi}{2} \tag{10.13}$$

where $\theta_m(0)$ is the initial rotor angle, measured with respect to the phase-a magnetic axis.

Example 10.1
In a three-phase, 2-pole, PMAC motor, the torque constant $k_T = 0.5$ Nm/A. Calculate the phase currents if the motor is to produce a counter-clockwise holding torque of 5 Nm to keep the rotor, which is at an angle of $\theta_m = 45°$, from turning.

Solution   From Equation 10.6, $\hat{I}_s = T_{em}/k_T = 10$ A. From Equation 10.10, $\theta_{i_s} = \theta_m + 90° = 135°$. Therefore, $\vec{i}_s(t) = \hat{I}_s \angle \theta_{i_s} = 10 \angle 135°$ A, as shown in Figure 10.7. From Equations 10.12a through 10.12c,

$$i_a = \frac{2}{3}\,\hat{I}_s \cos\theta_{i_s} = -4.71 \text{ A},$$

$$i_b = \frac{2}{3}\,\hat{I}_s \cos(\theta_{i_s} - 120°) = 6.44 \text{ A}, \text{ and}$$

$$i_c = \frac{2}{3}\,\hat{I}_s \cos(\theta_{i_s} - 240°) = -1.73 \text{ A}.$$

Since the rotor is not turning, the phase currents in this example are dc.

10.3.5   Induced EMFs in the Stator Windings During Balanced Sinusoidal Steady State

In the stator windings, emfs are induced due to two flux-density distributions:

1. As the rotor rotates with an instantaneous speed of $\omega_m(t)$, so does the space vector $\vec{B}_r(t)$ shown in Figure 10.3a. This rotating flux-density distribution "cuts" the stator windings to induce a back-emf in them.

2. The stator phase-winding currents under a balanced sinusoidal steady state produce a rotating flux-density distribution due to the rotating $\vec{i}_s(t)$ space vector. This rotating flux-density distribution induces emfs in the stator windings, similar to those induced by the magnetizing currents in the previous chapter.

Neglecting saturation in the magnetic circuit, the emfs that were induced, due to the two causes mentioned above, can be superimposed to calculate the resultant emf in the stator windings. In the following subsections, we will assume a 2-pole machine in a balanced sinusoidal steady state, with a rotor speed of ω_m in the counter-clockwise direction. We will also assume that at $t = 0$ the rotor is at $\theta_m = -90°$ for ease of drawing the space vectors.

10.3.5.1 Induced EMF in the Stator Windings due to Rotating $\vec{B}_r(t)$

We can make use of the analysis in the previous chapter that led to Equation 9.41. In the present case, the rotor flux-density vector $\vec{B}_r(t)$ is rotating at the instantaneous speed of ω_m with respect to the stator windings. Therefore, in Equation 9.41, substituting $\vec{B}_r(t)$ for $\vec{B}_{ms}(t)$ and ω_m for ω_{syn},

$$\vec{e}_{ms,\vec{B}_r}(t) = j\omega_m \frac{3}{2}\left(\pi\, r\, \ell \frac{N_s}{2}\right)\vec{B}_r(t) \tag{10.14}$$

We can define a voltage constant k_E, equal to the torque constant k_T in Equation 10.6 for a 2-pole machine:

$$k_E\left[\frac{V}{rad/s}\right] = k_T\left[\frac{Nm}{A}\right] = \pi\, r\, \ell\, \frac{N_s}{2}\, \hat{B}_r \tag{10.15}$$

where $\hat{B}_r$ (the peak of the rotor-produced flux-density) is a constant in permanent-magnet synchronous motors. In terms of the voltage constant k_E, the induced voltage space vector in Equation 10.14 can be written as

$$\vec{e}_{ms,\vec{B}_r}(t) = j\frac{3}{2}k_E\omega_m \angle \theta_m(t) = \frac{3}{2}k_E\omega_m \angle \{\theta_m(t) + 90°\} \tag{10.16}$$

The rotor flux-density space vector $\vec{B}_r(t)$ and the induced-emf space vector $\vec{e}_{ms,\vec{B}_r}(t)$ are drawn for time $t = 0$ in Figure 10.8a.

10.3.5.2 Induced EMF in the Stator Windings due to Rotating $\vec{i}_s(t)$: Armature Reaction

In addition to the flux-density distribution in the air gap created by the rotor magnets, another flux-density distribution is established by the stator phase currents. As shown in Figure 10.8b, the stator-current space vector $\vec{i}_s(t)$ at time $t = 0$ is made to lead the rotor position by 90°. Because we are operating under a balanced sinusoidal steady state, we can make use of the analysis in the previous chapter, where Equation 9.40 showed the relationship between the induced-emf space vector and the stator-current space vector.

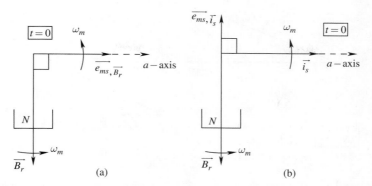

FIGURE 10.8 (a) Induced emf due to rotating rotor flux density space vector; (b) Induced emf due to rotating stator current space vector.

Thus, in the present case, due to the rotation of $\vec{i}_s(t)$, the induced voltages in the stator phase windings can be represented as

$$\vec{e_{ms,\,i_s}}(t) = j\omega_m L_m \vec{i}_s(t) \tag{10.17}$$

Space vectors $\vec{e_{ms,\,i_s}}$ and $\vec{i}_s$ are shown in Figure 10.8b at time $t = 0$.

Note that the magnetizing inductance L_m in the PMAC motor has the same meaning as in the generic ac motors discussed in Chapter 9. However, in PMAC motors, the rotor on its surface has permanent magnets (exceptions are motors with interior permanent magnets) whose permeability is effectively that of the air gap. Therefore, PMAC motors have a larger equivalent air gap, thus resulting in a smaller value of L_m (see Equation 9.36 of the previous chapter).

10.3.5.3 Superposition of the Induced EMFs in the Stator Windings

In PMAC motors, rotating $\vec{B}_r(t)$ and $\vec{i}_s(t)$ are present simultaneously. Therefore, the emfs induced due to each one can be superimposed (assuming no magnetic saturations) to obtain the resultant emf (excluding the leakage flux of the stator windings):

$$\vec{e_{ms}}(t) = \vec{e_{ms,\,B_r}}(t) + \vec{e_{ms,\,i_s}}(t) \tag{10.18}$$

Substituting from Equations 10.16 and 10.17 into Equation 10.18, the resultant induced emf $\vec{e_{ms}}(t)$ is

$$\vec{e_{ms}}(t) = \frac{3}{2} k_E \omega_m \angle \{\theta_m(t) + 90°\} + j\omega_m L_m \vec{i}_s(t) \tag{10.19}$$

The space vector diagram is shown in Figure 10.9a at time $t = 0$. The phase-*a* phasor equation corresponding to the space vector equation above can be written, noting that the phasor amplitudes are smaller than the space vector amplitudes by a factor of 3/2, but the phasor and the corresponding space vector have the same orientation:

$$\bar{E}_{ma} = \underbrace{k_E \omega_m \angle \{\theta_m(t) + 90°\}}_{\bar{E}_{ma,\,\vec{B_r}}} + j\omega_m L_m \bar{I}_a \tag{10.20}$$

The phasor diagram from Equation 10.20 for phase-*a* is shown in Figure 10.9b.

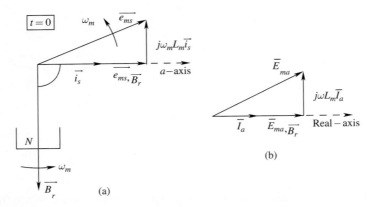

FIGURE 10.9 (a) Space vector diagram of induced emfs; (b) Phasor diagram for phase-a.

10.3.5.4 Per-Phase Equivalent Circuit

Corresponding to the phasor representation in Equation 10.20 and the phasor diagram in Figure 10.9b, a per-phase equivalent circuit for phase-a can be drawn as shown in Figure 10.10a. The voltage $\overline{E}_{ma,\vec{B}_r}$, induced due to the rotation of the rotor field distribution $\vec{B}_r$, is represented as an induced back-emf. The second term on the right side of Equation 10.20 is represented as a voltage drop across the magnetizing inductance L_m. To complete this per-phase equivalent circuit, the stator-winding leakage inductance $L_{\ell s}$ and the resistance R_s are added in series. The sum of the magnetizing inductance L_m and the leakage inductance $L_{\ell s}$ is called the synchronous inductance L_s:

$$L_s = L_{\ell s} + L_m \tag{10.21}$$

We can simplify the equivalent circuit of Figure 10.10a by neglecting the resistance and by representing the two inductances by their sum, L_s, as done in Figure 10.10b. To simplify the notation, the induced back-emf is called the field-induced back-emf $\overline{E}_{fa}$ in phase-a, where, from Equation 10.20, the peak of this voltage in each phase is

$$\hat{E}_f = k_E \omega_m \tag{10.22}$$

Notice that in PMAC drives the power-processing unit is a source of controlled currents such that $\overline{I}_a$ is in phase with the field-induced back-emf $\overline{E}_{fa}$, as confirmed by the phasor diagram of Figure 10.9b. The power-processing unit supplies this current by producing a voltage that, for phase-a in Figure 10.10b is

$$\overline{V}_a = \overline{E}_{fa} + j\omega_m L_s \overline{I}_a \tag{10.23}$$

Example 10.2

In a 2-pole, three-phase PMAC drive, the torque constant k_T and the voltage constant k_E are 0.5 in *MKS* units. The synchronous inductance is *15* mH (neglect the winding resistance). This motor is supplying a torque of *3* Nm at a speed of *3,000* rpm in a balanced sinusoidal steady state. Calculate the per-phase voltage across the power-processing unit as it supplies controlled currents to this motor.

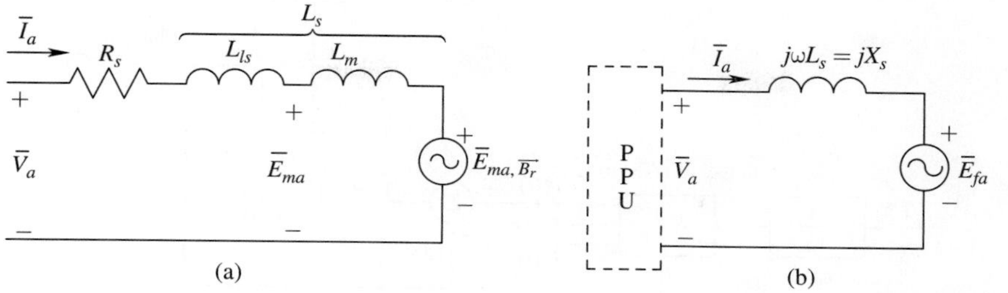

FIGURE 10.10 (a) Per-phase equivalent circuit; (b) simplified equivalent circuit.

Solution From Equation 10.6, $\hat{I}_s = \frac{3.0}{0.5} = 6$ A, and $\hat{I}_a = \frac{2}{3}\hat{I}_s = 4$ A. The speed $\omega_m = \frac{3000}{60}(2\pi) = 314.16$ rad/s. From Equation 10.22, $\hat{E}_f = k_E\omega_m = 0.5 \times 314.16 = 157.08$ V.

Assuming $\theta_m(0) = -90°$, from Equation 10.10, $\theta_{i_s}|_{t=0} = 0°$. Hence, in the per-phase equivalent circuit of Figure 10.10b, $\overline{I}_a = 4.0\angle 0°$ A and $\overline{E}_{fa} = 157.08\angle 0°$ V. Therefore, from Equation 10.23, in the per-phase equivalent circuit of Figure 10.10b,

$$\overline{V}_a = \overline{E}_{fa} + j\omega_m L_s \overline{I}_a = 157.08\angle 0° + j314.16 \times 15 \times 10^{-3} \times 4.0\angle 0° = 157.08 + j18.85$$
$$= 158.2 \angle 6.84° \text{ V}.$$

10.3.6 Generator-Mode of Operation of PMAC Drives

PMAC drives can operate in their generator mode simply by controlling the stator-current space vector $\vec{i}_s$ to be 90° behind the $\vec{B}_r$ space vector in Figure 10.3a in the direction of rotation. This will result in current directions in the conductors of the hypothetical winding to be opposite of what is shown in Figure 10.3b. Hence, the electromagnetic torque produced will be in a direction to oppose the torque supplied by the prime-mover that is causing the rotor to rotate. An analysis similar to the motoring-mode can be carried out in the generator-mode of operation.

10.4 THE CONTROLLER AND THE POWER-PROCESSING UNIT (PPU)

As shown in the block diagram of Figure 10.1, the task of the controller is to dictate the switching in the power-processing unit, such that the desired currents are supplied to the PMAC motors. This is further illustrated in Figure 10.11a, where phases b and c are omitted for simplification. The reference signal T_{em}^* is generated from the outer speed and position loops discussed in Chapter 8. The rotor position θ_m is measured by the resolver (discussed in Chapter 17) connected to the shaft. Knowing the torque constant k_T allows us to calculate the reference current $\hat{I}_s^*$ to be T_{em}^*/k_T (from Equation 10.9). Knowing $\hat{I}_s^*$ and θ_m allows the reference currents i_a^*, i_b^*, and i_c^* to be calculated at any instant of time from Equation 10.11 and Equations 10.12a through 10.12c.

One of the easiest ways to ensure that the motor is supplied the desired currents is to use hysteresis control similar to that discussed in Chapter 7 for ECM drives. The measured phase current is compared with its reference value in the hysteresis comparator,

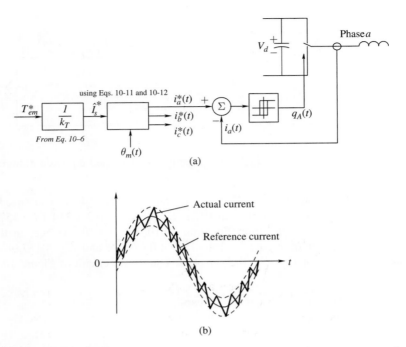

FIGURE 10.11 (a) Block diagram representation of hysteresis current control; (b) Current waveform.

whose output determines the switch state (up or down), resulting in current as shown in Figure 10.11b.

In spite of the simplicity of the hysteresis control, one perceived drawback of this controller is that the switching frequency changes as a function of the back-emf waveform. For this reason, constant switching frequency controllers are used. They are beyond the scope of this book, but Reference [3], listed at the end of this chapter, is an excellent source of information on them.

10.5 LOAD-COMMUTATED-INVERTER (LCI) SUPPLIED SYNCHRONOUS MOTOR DRIVES

Applications such as induced-draft fans and boiler feed-water pumps in the central power plants of electric utilities require adjustable-speed drives in very large power ratings, often in excess of one megawatt. At these power levels, even the slightly higher efficiency of synchronous motors, compared to the induction motors that we will discuss in the next three chapters, can be substantial. Moreover, to adjust the speed of synchronous motors, it is possible to use thyristor-based power-processing units, which are less expensive at these megawatt power ratings compared to the switch-mode power-processing units discussed in Chapter 4.

The block diagram of LCI drives is shown in Figure 10.12, where the synchronous motor has a field winding on the rotor, which is supplied by a dc current that can be adjusted, thus providing another degree of control.

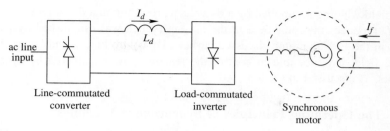

FIGURE 10.12   LCI-Synchronous motor drive.

On the utility side, a line-commutated thyristor converter, which is described in Chapter 17 in connection with dc drives, is used. A similar converter is used on the motor side, where the commutation of currents is provided by the load, which in this case is a synchronous machine. This is also the reason for calling the motor-side converter a load-commutated inverter (LCI). A filter inductor, which makes the input to the load-commutated inverter appear as a dc current source, is used in the dc-link between the two converters. Hence, this inverter is also called a current-source inverter (in contrast to the switch-mode converters discussed in Chapter 4, where a parallel-connected capacitor appears as a dc voltage source—thus such converters are sometimes called voltage-source inverters). Further details of LCI-synchronous motor drives are in Reference [1].

10.6   SYNCHRONOUS GENERATORS

Today it is rare for synchronous machines without any power electronics interface (PPU) to be used as motors, which run at a constant speed that is dictated by the frequency of the utility grid. In the past, constant-speed synchronous machines in large power ratings were used as synchronous condensers (many still exist) in utility substations to provide voltage support and stability enhancement. However, the recent trend is to use static (semiconductor-based) controllers, which can provide reactive power (leading and lagging) without the maintenance problems associated with rotating equipment. Therefore, the role of synchronous machines is mainly to generate electricity in large central power plants of electric utilities, where they are driven by turbines fueled by gas, by steam in coal-fired or nuclear plants, or propelled by water flow in hydroelectric plants.

10.6.1   The Structure of Synchronous Machines

In the above application, turbines and synchronous generators are large and massive, but their stator windings, in principle, are the same as their smaller-power counterpart. Generators driven by gas and steam turbines often rotate at high speeds and thus have a 2-pole, round-rotor structure. Hydraulic-turbine–driven generators operate at very low speeds, and thus must have a large number of poles to generate a 60-Hz (or 50-Hz) frequency. This requires a salient-pole structure for the rotor, as discussed in Chapter 6. This saliency causes unequal magnetic reluctance along various paths through the rotor. Analysis of such salient-pole machines requires a sophisticated analysis, which is beyond the scope of this book. Therefore, we will assume the rotor to be perfectly round (non-salient) with a uniform air gap and thus to have a uniform reluctance in the path of flux lines.

A field winding is supplied by a dc voltage, resulting in a dc current I_f. The field-current I_f produces the rotor field in the air gap (which was established by permanent magnets in the PMAC motor discussed earlier). By controlling I_f and hence the rotor-produced field, it is possible to control the reactive power delivered by synchronous generators, as discussed in section 10.6.2.2.

10.6.2 The Operating Principles of Synchronous Machines

In steady state, the synchronous generator must rotate at the synchronous speed estab-lished by the line-fed stator windings. Therefore, in steady state, the per-phase equivalent circuit of PMAC motor drives in Figure 10.10a or 10.10b applies to the synchronous machines as well. The important difference is that in PMAC motor drives a PPU is present, which under feedback of rotor position supplies appropriate phase currents to the motor. Of course, the PPU produces a voltage $\overline{V}_a$ shown in Figure 10.10b, but its main purpose is to supply controlled currents to the motor.

Line-connected synchronous machines lack the control over the currents that PMAC drives have. Rather, on a per-phase basis, synchronous machines have two voltage sources, as shown in Figure 10.13a—one belonging to the utility source and the other to the internally-induced back-emf $\overline{E}_{fa}$. Following the generator convention, the current is defined as being supplied by the synchronous generator, as shown in Figure 10.13a. This current can be calculated as follows where $\overline{V}_a$ is chosen as the reference phasor $(\overline{V}_a = \hat{V} \angle 0°)$ and the torque angle δ associated with $\overline{E}_{fa}$ is positive in the gen-erator mode:

$$\overline{I}_a = \frac{\overline{E}_{fa} - \overline{V}_a}{jX_s} = \frac{\hat{E}_f \sin \delta}{X_s} - j\frac{\hat{E}_f \cos \delta - \hat{V}}{X_s} \tag{10.24}$$

Taking the conjugate of $\overline{I}_a$ (represented by "*" as a superscript),

$$\overline{I}_a^* = \frac{\hat{E}_f \sin \delta}{X_s} + j\frac{\hat{E}_f \cos \delta - \hat{V}}{X_s} \tag{10.25}$$

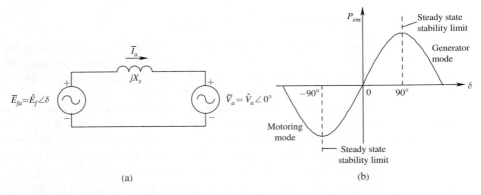

FIGURE 10.13 (a) Synchronous generator; (b) Power angle characteristics.

The total (three-phase) power supplied by the generator, in terms of peak quantities, is

$$P_{em} = \frac{3}{2}\mathrm{Re}(\overline{V}_a \overline{I}_a^*) = \frac{3}{2}\hat{V}\mathrm{Re}\left[\frac{\hat{E}_f \sin\delta}{X_s} + j\frac{\hat{E}_f \cos\delta - \hat{V}}{X_s}\right] \quad \text{or}$$

$$P_{em} = \frac{3}{2}\frac{\hat{E}_f \hat{V} \sin\delta}{X_s} \tag{10.26}$$

If the field current is constant, $\hat{E}_f$ at the synchronous speed is also constant, and thus the power output of the generator is proportional to the sine of the torque angle δ between $\overline{E}_{fa}$ and $\overline{V}_a$. This power-angle relationship is plotted in Figure 10.13b for both positive and negative values of δ.

We should note that the power and torque associated with a machine are proportional to each other, related by the rotor speed, which is constant in steady state. Therefore in steady state, the toque angle δ is synonymous as the power angle. This torque (power) angle δ is the angle between the internally-induced back-emf and the terminal voltage (assuming the stator resistance to be zero) in the per-phase equivalent circuit of Figure 10.13a for a synchronous machine. In PMAC drives, to relate them to synchronous machines, this angle δ is the angle between $\vec{e}_{ms}$ and $\vec{e}_{ms,B_r}$ in Figure 10.9a of PMAC drives, where again the torque (power) output is proportional to $\sin\delta$. In both cases, this angle δ is the angle between the voltages induced in the stator windings by the rotor flux and by the resultant flux (superposition of the rotor flux and that due to stator currents).

10.6.2.1 Stability and Loss of Synchronism

Figure 10.13b shows that the power supplied by the synchronous generator, as a function of δ, reaches its peak at 90°. This is the steady-state limit, beyond which the synchronism is lost. This can be explained as follows: for values of δ below 90 degrees, to supply more power, the power input from the mechanical prime-mover is increased (for example, by letting more steam into the turbine). This momentarily speeds up the rotor, causing the torque angle δ associated with the rotor-induced voltage $\overline{E}_{fa}$ to increase. This in turn, from Equation 10.26, increases the electrical power output, which finally settles at a new steady state with a higher value of the torque angle δ. However, beyond $\delta = 90$ degrees, increasing δ causes the output power to decline, which results in a further increase in δ (because more mechanical power is coming in while less electrical power is going out). This increase in δ causes an intolerable increase in machine currents, and the circuit breakers trip to isolate the machine from the grid, thus saving the machine from being damaged.

The above sequence of events is called the "loss of synchronism," and the stability is lost. In practice, transient stability, in which there may be a sudden change in the electrical power output, forces the maximum value of the steady-state torque angle δ to be much less than 90 degrees, typically in a range of 30 to 45 degrees. A similar explanation applies to the motoring mode with negative values of δ.

10.6.2.2 Field (Excitation) Control to Adjust Reactive Power and Power Factor

The reactive power associated with synchronous machines can be controlled in magnitude as well as in sign (leading or lagging). To discuss this, let us assume, as a base case,

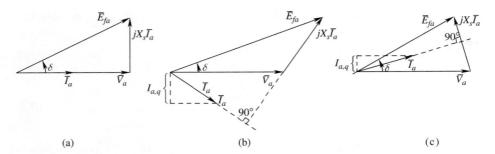

FIGURE 10.14   (a) Synchronous generator at (a) Unity power factor; (b) Over-excited; and (c) Under-excited.

that a synchronous generator is supplying a constant power, and the field current I_f is adjusted such that this power is supplied at a unity power factor, as shown in the phasor diagram of Figure 10.14a.

Over-Excitation: Now, an increase in the field current (called over-excitation) will result in a larger magnitude of $\overline{E}_{fa}$ (assuming no magnetic saturation, $\hat{E}_f$ depends linearly on the field current I_f). However, $\hat{E}_f \sin\delta$ must remain constant (from Equation 10.26, since the power output is constant). This results in the phasor diagram of Figure 10.14b, where the current is lagging $\overline{V}_a$. Considering the utility grid to be a load (which it is, in the generator mode of the machine), it absorbs reactive power as an inductive load does. Therefore, the synchronous generator, operating in an over-excited mode, supplies reactive power the way a capacitor does. The three-phase reactive power Q can be computed from the reactive component of the current $I_{a,q}$ as

$$Q = \frac{3}{2}\hat{V}I_{a,q} \tag{10.27}$$

Under-Excitation: In contrast to over-excitation, decreasing I_f results in a smaller magnitude $\hat{E}_f$, and the corresponding phasor diagram, assuming that the power output remains constant as before, can be represented as in Figure 10.14c. Now the current $\overline{I}_a$ leads the voltage $\overline{V}_a$, and the load (the utility grid) supplies reactive power as a capacitive load does. Thus, the generator in an under-excited mode absorbs reactive power the way an inductor does.

Similar control over the reactive power can be observed by drawing phasor diagrams, if the machine is operating as a synchronous motor (see homework problem 10.10). The reactive power of the machine can be calculated, similar to the calculations of the real power that led to Equation 10.26. This is left as homework problem 10.11.

SUMMARY/REVIEW QUESTIONS

1. List various names associated with the PMAC drives and the reasons behind them.
2. Draw the overall block diagram of a PMAC drive. Why must they operate in a closed-loop?
3. How do sinusoidal PMAC drives differ from the ECM drives described in Chapter 7?

4. Ideally, what are the flux-density distributions produced by the rotor and the stator phase windings?
5. What does the $\vec{B}_r(t) = \hat{B}_r \angle \theta_m(t)$ space vector represent?
6. In PMAC drives, why at all times is the $\vec{i}_s(t)$ space vector placed 90 degrees ahead of the $\vec{B}_r(t)$ space vector in the intended direction of rotation?
7. Why do we need to measure the rotor position in PMAC drives?
8. What does the electromagnetic torque produced by a PMAC drive depend on?
9. How can regenerative braking be accomplished in PMAC drives?
10. Why are PMAC drives called self-synchronous? How is the frequency of the applied voltages and currents determined? Are they related to the rotational speed of the shaft?
11. In a p-pole PMAC machine, what is the angle of the $\vec{i}_s(t)$ space vector in relation to the phase-a axis, for a given θ_m?
12. What is the frequency of currents and voltages in the stator circuit needed to produce a holding torque in a PMAC drive?
13. In calculating the voltages induced in the stator windings of a PMAC motor, what are the two components that are superimposed? Describe the procedure and the expressions.
14. Does L_m in the per-phase equivalent circuit of a PMAC machine have the same expression as in Chapter 9? Describe the differences, if any.
15. Draw the per-phase equivalent circuit and describe its various elements in PMAC drives.
16. Draw the controller block diagram, and describe the hysteresis control of PMAC drives.
17. What is an LCI-synchronous motor drive? Describe it briefly.
18. For what purpose are line-connected synchronous generators used?
19. Why are there problems of stability and loss of synchronism associated with line-connected synchronous machines?
20. How can the power factor associated with synchronous generators be made to be leading or lagging?

REFERENCES

1. N. Mohan, *Power Electronics: A First Course* (New York: John Wiley & Sons, 2011).
2. N. Mohan, T. Undeland, and W. Robbins, *Power Electronics: Converters, Applications, and Design*, 2nd ed. (New York: John Wiley & Sons, 1995).
3. T. Jahns, *Variable Frequency Permanent Magnet AC Machine Drives, Power Electronics and Variable Frequency Drives,* edited by B. K. Bose (IEEE Press, 1997).
4. M. P. Kazmierkowski and H. Tunia, *Automatic Control of Converter-Fed Drives* (Amsterdam: Elsevier, 1994).

PROBLEMS

10.1 Calculate the torque constant, similar to that in Equation 10.6, for a 4-pole machine, where N_s equals the total number of turns per-phase.
10.2 Prove that Equation 10.11 is correct.
10.3 Repeat Example 10.1 for $\theta_m = -45°$.

10.4 Repeat Example 10.1 for a 4-pole machine with the same value of k_T as in Example 10.1.

10.5 The PMAC machine of Example 10.2 is supplying a load torque $T_L = 5$ Nm at a speed of 5,000 rpm. Draw a phasor diagram showing $\overline{V}_a$ and $\overline{I}_a$, along with their calculated values.

10.6 Repeat Problem 10.5 if the machine has $p = 4$, but has the same values of k_E, k_T, and L_s as before.

10.7 Repeat Problem 10.5, assuming that at time t = 0, the rotor angle $\theta_m(0) = 0°$.

10.8 The PMAC motor in Example 10.2 is driving a purely inertial load. A constant torque of 5 Nm is developed to bring the system from rest to a speed of 5,000 rpm in 5 s. Neglect the stator resistance and the leakage inductance. Determine and plot the voltage $v_a(t)$ and the current $i_a(t)$ as functions of time during this 5-second interval.

10.9 In Problem 10.8, the drive is expected to go into a regenerative mode at $t = 5^+$ s, with a torque $T_{em} = -5$ Nm. Assume that the rotor position at this instant is zero: $\theta_m = 0$. Calculate the three stator currents at this instant.

10.10 Redraw Figure 10.3 if the PMAC drive is operating as a generator.

10.11 Recalculate Example 10.2 if the PMAC drive is operating as a generator, and instead of it supplying a torque of 37 Nm, it is being supplied this torque from the mechanical system connected to the machine-shaft.

10.12 The 2-pole motor in a PMAC drive has the following parameters: $R_s = 0.416\ \Omega$, $L_s = 1.365$ mH, and $k_T = 0.0957$ Nm/A. Draw the space vector and phasor diagrams for this machine if it is supplying its continuous rated torque of $T_{em} = 3.2$ Nm at its rated speed of 6,000 rpm.

10.13 Draw the phasor diagrams associated with under-excited and over-excited synchronous motors and show the power factor of operation associated with each.

10.14 Calculate the expression for the reactive power in a 3-phase synchronous machine in terms of $\hat{E}_f$, $\hat{V}$, X_s, and δ. Discuss the influence of $\hat{E}_f$.

11

INDUCTION MOTORS: BALANCED, SINUSOIDAL STEADY STATE OPERATION

11.1 INTRODUCTION

Induction motors with squirrel-cage rotors are the workhorses of industry because of their low cost and rugged construction. When operated directly from line voltages (a 50- or 60-Hz utility input at essentially a constant voltage), induction motors operate at a nearly constant speed. However, by means of power electronic converters, it is possible to vary their speed efficiently. Induction-motor drives can be classified into two broad categories based on their applications:

1. *Adjustable-Speed Drives.* An important application of these drives is to adjust the speeds of fans, compressors, pumps, blowers, and the like in the process control industry. In a large number of applications, this capability to vary speed efficiently can lead to large savings. Adjustable-speed induction-motor drives are also used for electric traction, including hybrid vehicles.
2. *Servo Drives.* By means of sophisticated control discussed in Chapter 12, induction motors can be used as servo drives in machine tools, robotics, and so on by emulating the performance of dc-motor drives and brushless-dc motor drives.

Because the subject of induction-motor drives is extensive, we will cover it in three separate chapters. In this chapter, we will examine the behavior of three-phase induction machines supplied by balanced sinusoidal line-frequency voltages at their rated values. In Chapter 12, we will discuss energy-efficient speed control of induction motor drives for process control and traction applications. In Chapter 13, we will discuss single-phase induction motors, universal motors, and their speed control.

There are many varieties of induction motors. Single-phase induction motors are used in low power ratings (fractional kW to a few kW), as discussed in Chapter 13, in applications where their speed does not have to be controlled in a continuous manner. Wound-rotor induction generators are used in large power ratings (300 kW and higher) for wind-electric generation. However, our focus in this chapter and in subsequent ones is on three-phase, squirrel-cage induction motors, which are commonly used in adjustable-speed applications.

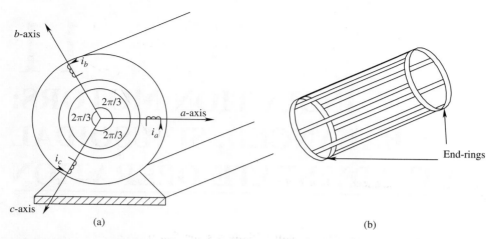

FIGURE 11.1   (a) Three-phase stator winding axes; (b) squirrel-cage rotor.

11.2   THE STRUCTURE OF THREE-PHASE, SQUIRREL-CAGE INDUCTION MOTORS

The stator of an induction motor consists of three-phase windings, sinusoidally-distributed in the stator slots as discussed in Chapter 9. These three windings are displaced by 120° in space with respect to each other, as shown by their axes in Figure 11.1a.

The rotor, consisting of a stack of insulated laminations, has electrically conducting bars of copper or aluminum inserted (molded) through it, close to the periphery in the axial direction. These bars are electrically shorted at each end of the rotor by electrically-conducting end-rings, thus producing a cage-like structure, as shown in Figure 11.1b. Such a rotor, called a squirrel-cage rotor, has a simple construction, low-cost, and rugged nature.

11.3   THE PRINCIPLES OF INDUCTION MOTOR OPERATION

Our analysis will be under the line-fed conditions in which a balanced set of sinusoidal voltages of rated amplitude and frequency are applied to the stator windings. In the following discussion, we will assume a 2-pole structure that can be extended to a multi-pole machine with p > 2.

Figure 11.2 shows the stator windings. Under a balanced sinusoidal steady state condition, the motor-neutral "n" is at the same potential as the source-neutral. Therefore, the source voltages v_a and so on appear across the respective phase windings, as shown in Figure 11.2a. These phase voltages are shown in the phasor diagram of Figure 11.2b, where

$$\overline{V}_a = \hat{V} \angle 0°, \quad \overline{V}_b = \hat{V} \angle -120°, \quad \text{and} \quad \overline{V}_c = \hat{V} \angle -240° \tag{11.1}$$

and $f\left(= \frac{\omega}{2\pi}\right)$ is the frequency of the applied line-voltages to the motor.

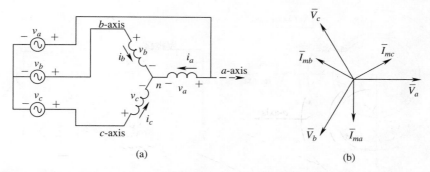

FIGURE 11.2 Balanced 3-phase sinusoidal voltages applied to the stator, rotor open-circuited.

To simplify our analysis, we will initially assume that the stator windings have a zero resistance ($R_s = 0$). Also, we will assume that $L_{\ell s} = 0$, implying that the leakage flux is zero; that is, all of the flux produced by each stator winding crosses the air gap and links with the other two stator windings and the rotor.

11.3.1 Electrically Open-Circuited Rotor

Initially, we will assume that the rotor is magnetically present but that its rotor bars are somehow open-circuited so that no current can flow. Therefore, we can use the analysis of Chapter 9, where the applied stator voltages given in Equation 11.1 result only in the following magnetizing currents, which establish the rotating flux-density distribution in the air gap:

$$\overline{I}_{ma} = \hat{I}_m \angle -90°, \quad \overline{I}_{mb} = \hat{I}_m \angle -210°, \quad \text{and} \quad \overline{I}_{mc} = \hat{I}_m \angle -330° \tag{11.2}$$

These phasors are shown in Figure 11.2b, where, in terms of the per-phase magnetizing inductance L_m, the amplitude of the magnetizing currents is

$$\hat{I}_m = \frac{\hat{V}}{\omega L_m} \tag{11.3}$$

The space vectors at $t=0$ are shown in Figure 11.3a, where, from Chapter 9,

$$\vec{v}_s(t) = \frac{3}{2} \hat{V} \angle \omega t \tag{11.4}$$

$$\vec{i}_{ms}(t) = \frac{3}{2} \hat{I}_m \angle \left(\omega t - \frac{\pi}{2}\right) \tag{11.5}$$

$$\vec{B}_{ms}(t) = \frac{\mu_o N_s}{2\ell_g} \hat{I}_{ms} \angle \left(\omega t - \frac{\pi}{2}\right) \quad \text{where} \quad \hat{B}_{ms} = \frac{3}{2} \frac{\mu_o N_s}{2\ell_g} \hat{I}_m \quad \text{and} \tag{11.6}$$

$$\vec{v}_s(t) = \vec{e}_{ms}(t) = j\omega \left(\frac{3}{2} \pi r \ell \frac{N_s}{2}\right) \vec{B}_{ms}(t) \tag{11.7}$$

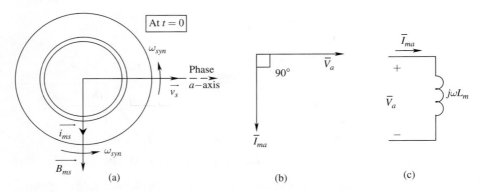

FIGURE 11.3 Space vector representations at time t=0; (b) voltage and current phasors for phase-*a*; (c) equivalent circuit for phase-*a*.

These space vectors rotate at a constant synchronous speed ω_{syn}, which in a 2-pole machine is

$$\omega_{syn} = \omega \quad \left(\omega_{syn} = \frac{\omega}{p/2} \text{ for a } p\text{-pole machine} \right) \tag{11.8}$$

Example 11.1

A 2-pole, 3-phase induction motor has the following physical dimensions: radius $r = 7$ cm, length $\ell = 9$ cm, and the air gap length $\ell_g = 0.5$ mm. Calculate N_s, the number of turns per-phase, so that the peak of the flux-density distribution does not exceed 0.8 T when the rated voltages of 208 V (line-line, rms) are applied at the 60-Hz frequency.

Solution From Equation 11.7, the peak of the stator voltage and the flux-density distribution space vectors are related as follows:

$$\hat{V}_s = \frac{3}{2}\pi r\ell \, \frac{N_s}{2} \omega \hat{B}_{ms} \quad \text{where} \quad \hat{V}_s = \frac{3}{2}\hat{V} = \frac{3}{2}\frac{208\sqrt{2}}{\sqrt{3}} = 254.75 \text{ V}$$

Substituting given values in the above expression, $N_s = 56.9$ turns. Since the number of turns must be an integer, $N_s \simeq 57$ turns is selected.

11.3.2 The Short-Circuited Rotor

The voltages applied to the stator completely dictate the magnetizing currents (see Equations 11.2 and 11.3) and the flux-density distribution, which is represented in Equation 11.6 by $\vec{B}_{ms}(t)$ and is "cutting" the stator windings. Assuming the stator winding resistances and the leakage inductances to be zero, this flux-density distribution is unaffected by the currents in the rotor circuit, as illustrated by the transformer analogy below.

Transformer Analogy: A two-winding transformer is shown in Figure 11.4a, where two air gaps are introduced to bring the analogy closer to the case of induction machines where flux lines must cross the air gap twice. The primary winding resistance and the

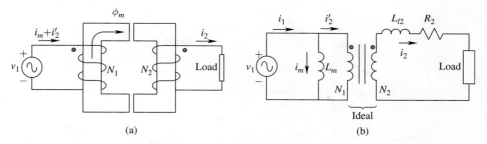

(a)   (b)

FIGURE 11.4 (a) Two winding transformer; (b) equivalent circuit of the Two winding transformer.

leakage inductance are neglected (similar to neglecting the stator winding resistances and leakage inductances). The transformer equivalent circuit is shown in Figure 11.4b. The applied voltage $v_1(t)$ and the flux $\phi_m(t)$ linking the primary winding are related by Faraday's Law:

$$v_1 = N_1 \frac{d\phi_m}{dt} \tag{11.9}$$

or, in the integral form,

$$\phi_m(t) = \frac{1}{N_1} \int v_1 \cdot dt \tag{11.10}$$

This shows that in this transformer, the flux $\phi_m(t)$ linking the primary winding is completely determined by the time-integral of $v_1(t)$, independent of the current i_2 in the secondary winding.

This observation is confirmed by the transformer equivalent circuit of Figure 11.4b, where the magnetizing current i_m is completely dictated by the time-integral of $v_1(t)$, independent of the currents i_2 and i_2':

$$i_m(t) = \frac{1}{L_m} \int v_1 \cdot dt \tag{11.11}$$

In the ideal transformer portion of Figure 11.4b, the ampere-turns produced by the load current $i_2(t)$ are "nullified" by the additional current $i_2'(t)$ drawn by the primary winding, such that

$$N_1 i_2'(t) = N_2 i_2(t) \quad \text{or} \quad i_2'(t) = \frac{N_2}{N_1} i_2(t) \tag{11.12}$$

Thus, the total current drawn by the primary winding is

$$i_1(t) = i_m(t) + \underbrace{\frac{N_2}{N_1} i_2(t)}_{i_2'(t)} \tag{11.13}$$

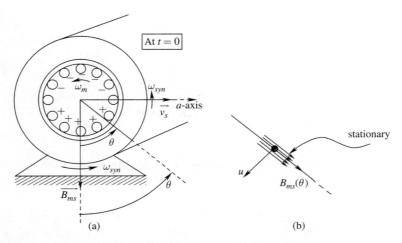

FIGURE 11.5 (a) Induced voltages in the rotor bar; (b) motion of the rotor bar relative to the flux density.

Returning to our discussion of induction machines, the rotor consists of a short-circuited cage made up of rotor bars and the two end-rings. Regardless of what happens in the rotor circuit, the flux-density distribution "cutting" the stator windings must remain the same as under the assumption of an open-circuited rotor, as represented by $\vec{B}_{ms}(t)$ in Equation 11.6.

Assume that the rotor is turning (due to the electromagnetic torque developed, as will be discussed shortly) at a speed ω_m in the same direction as the rotation of the space vectors, which represent the stator voltages and the air gap flux-density distribution. For now, we will assume that $\omega_m < \omega_{syn}$. The space vectors at time $t = 0$ are shown in the cross-section of Figure 11.5a.

There is a relative speed between the flux-density distribution rotating at ω_{syn} and the rotor conductors rotating at ω_m. This relative speed—that is, the speed at which the rotor is "slipping" with respect to the rotating flux-density distribution—is called the slip speed:

$$slip\ speed \quad \omega_{slip} = \omega_{syn} - \omega_m \tag{11.14}$$

By Faraday's Law ($e = B\ell u$), voltages are induced in the rotor bars due to the relative motion between the flux-density distribution and the rotor. At this time $t = 0$, the bar located at an angle θ from $\vec{B}_{ms}$ in Figure 11.5a is "cutting" a flux density $B_{ms}(\theta)$. The flux-density distribution is moving ahead of the bar at position θ at the angular speed of ω_{slip} rad/s or at the linear speed of $u = r\,\omega_{slip}$, where r is the radius. To determine the voltage induced in this rotor bar, we can consider the flux-density distribution to be stationary and the bar (at the angle θ) to be moving in the opposite direction at the speed u, as shown in Figure 11.5b. Therefore, the voltage induced in the bar can be expressed as

$$e_{bar}(\theta) = B_{ms}(\theta)\ell\underbrace{r\omega_{slip}}_{u} \tag{11.15}$$

where the bar is of length ℓ and is at a radius r. The direction of the induced voltage can be established by visualizing that on a positive charge q in the bar, the force f_q equals

$\mathbf{u} \times \mathbf{B}$ where $\mathbf{u}$ and $\mathbf{B}$ are vectors shown in Figure 11.5b. This force will cause the positive charge to move towards the front-end of the bar, establishing that the front-end of the bar will have a positive potential with respect to the back-end, as shown in Figure 11.5a.

At any time, the flux-density distribution varies as the cosine of the angle θ from its positive peak. Therefore in Equation 11.15, $B_{ms}(\theta) = \hat{B}_{ms} \cos \theta$. Hence,

$$e_{bar}(\theta) = \ell r \omega_{slip} \hat{B}_{ms} \cos \theta \qquad (11.16)$$

11.3.2.1 The Assumption of Rotor Leakage $L'_{\ell r} = 0$

At this point, we will make another *extremely* important simplifying assumption, to be analyzed later in more detail. The assumption is that the rotor cage has no leakage inductance; that is, $L'_{\ell r} = 0$. This assumption implies that the rotor has no leakage flux and that all of the flux produced by the rotor-bar currents crosses the air gap and links (or "cuts") the stator windings. Another implication of this assumption is that, at any time, the current in each squirrel-cage bar, short-circuited at both ends by conducting end-rings, is inversely proportional to the bar resistance R_{bar}.

In Figure 11.6a at $t = 0$, the induced voltages are maximum in the top and the bottom rotor bars "cutting" the peak flux density. Elsewhere, induced voltages in rotor bars depend on $\cos \theta$, as given by Equation 11.16. The polarities of the induced voltages at the near-end of the bars are indicated in Figure 11.6a. Figure 11.6b shows the electrical equivalent circuit that corresponds to the cross-section of the rotor shown in Figure 11.6a. The size of the voltage source represents the magnitude of the voltage induced. Because of the symmetry in this circuit, it is easy to visualize that the two end-rings (assumed to have negligible resistances themselves) are at the same potential. Therefore, the rotor bar at an angle θ from the positive flux-density peak location has a current that is equal to the induced voltage divided by the bar resistance:

$$i_{bar}(\theta) = \frac{e_{bar}(\theta)}{R_{bar}} = \frac{\ell r \omega_{slip} \hat{B}_{ms} \cos \theta}{R_{bar}} \qquad \text{(using Equation 11.16)} \qquad (11.17)$$

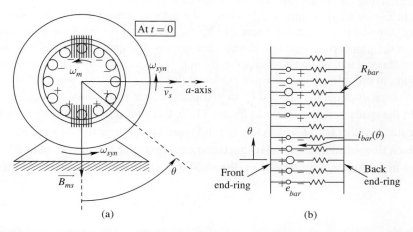

(a)   (b)

FIGURE 11.6 (a) Polarities of voltages induced; (b) electrical equivalent circuit of rotor.

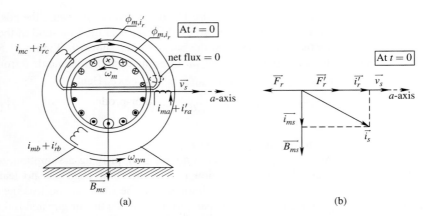

(a)   (b)

FIGURE 11.7   (a) Rotor-produced flux ϕ_{m,i_r} and the flux $\phi_{m,i_r'}$; (b) space vector diagram with short-circuited rotor ($L'_{\ell r} = 0$).

where each bar has a resistance R_{bar}. From Equation 11.17, the currents are maximum in the top and the bottom bars at this time, indicated by the largest circles in Figure 11.7a; elsewhere, the magnitude of the current depends on $\cos \theta$, where θ is the angular position of any bar as defined in Figure 11.6a.

It is important to note that the rotor has a uniform bar density around its periphery, as shown in Figure 11.6a. The sizes of the circles in Figure 11.7a denote the relative current magnitudes. The sinusoidal rotor-current distribution is different from that in the stator phase winding, which has a sinusoidally-distributed conductor density but the same current flowing through each conductor. In spite of this key difference, the outcome is the same—the ampere-turns need to be sinusoidally distributed in order to produce a sinusoidal field distribution in the air gap. In the rotor with a uniform bar density, a sinusoidal ampere-turn distribution is achieved because the currents in various rotor bars are sinusoidally distributed with position at any time.

The combined effect of the rotor-bar currents is to produce a sinusoidally-distributed mmf acting on the air gap. This mmf can be represented by a space vector $\vec{F}_r(t)$, as shown in Figure 11.7b at time $t = 0$. Due to the rotor-produced mmf, the resulting flux "cutting" the stator winding is represented by ϕ_{m,i_r} in Figure 11.7a. As argued earlier by means of the transformer analogy, the net flux-density distribution "cutting" the voltage-supplied stator windings must remain the same as in the case of the open-circuited rotor. Therefore, in order to cancel the rotor-produced flux ϕ_{m,i_r}, the stator windings must draw the additional currents i'_{ra}, i'_{rb}, and i'_{rc} to produce the flux represented by $\phi_{m,i_r'}$.

In the space vector diagram of Figure 11.7b, the mmf produced by the rotor bars is represented by $\vec{F}_r$ at time $t = 0$. As shown in Figure 11.7b, the stator currents i'_{ra}, i'_{rb}, and i_{rc}' (which flow in addition to the magnetizing currents) must produce an mmf $\vec{F}'_r$, which is equal in amplitude but opposite in direction to $\vec{F}_r$, in order to neutralize its effect:

$$\vec{F}'_r = -\vec{F}_r \tag{11.18}$$

The additional currents i'_{ra}, i'_{rb}, and i'_{rc} drawn by the stator windings to produce $\vec{F}'_r$ can be expressed by a current space vector $\vec{i}'_r$, as shown in Figure 11.7b at $t = 0$, where

$$\vec{i_r'} = \frac{\vec{F_r'}}{N_s/2} \quad (\hat{I}_r' = \text{the amplitude of } \vec{i_r'}) \tag{11.19}$$

The total stator current $\vec{i_s}$ is the vector sum of the two components: $\vec{i_{ms}}$, which sets up the magnetizing field, and $\vec{i_r'}$, which neutralizes the rotor-produced mmf:

$$\vec{i_s} = \vec{i_{ms}} + \vec{i_r'} \tag{11.20}$$

These space vectors are shown in Figure 11.7b at $t = 0$. Equation 11.17 shows that the rotor-bar currents are proportional to the flux-density peak and the slip speed. Therefore, the "nullifying" mmf peak and the peak current $\hat{I}_r'$ must also be linearly proportional to $\hat{B}_{ms}$ and ω_{slip}. This relationship can be expressed as

$$\hat{I}_r' = k_i \hat{B}_{ms} \omega_{slip} \quad (k_i = \text{a constant}) \tag{11.21}$$

where k_i is a constant based on the design of the machine.

During the sinusoidal steady state operating condition in Figure 11.7b, the rotor-produced mmf distribution (represented by $\vec{F_r}$) and the compensating mmf distribution (represented by $\vec{F_r'}$) rotate at the synchronous speed ω_{syn} and each has a constant amplitude. This can be illustrated by drawing the motor cross-section and space vectors at some arbitrary time $t_1 > 0$, as shown in Figure 11.8, where the $\vec{B_{ms}}$ space vector has rotated by an angle $\omega_{syn}t_1$ because $\vec{v_s}$ has rotated by $\omega_{syn}t_1$. Based on the voltages and currents induced in the rotor bars, $\vec{F_r}$ is still 90° behind the $\vec{B_{ms}}$ space vector, as in Figure 11.7a and 11.7b. This implies that the $\vec{F_r}(t)$ and $\vec{F_r'}(t)$ vectors are rotating at the same speed as $\vec{B_{ms}}(t)$—that is, the synchronous speed ω_{syn}. At a given operating condition with constant values of ω_{slip} and $\hat{B}_{ms}$, the bar-current distribution, relative to the peak of the flux-density vector, is the same in Figure 11.8 as in Figure 11.7. Therefore, the amplitudes of $\vec{F_r}(t)$ and $\vec{F_r'}(t)$ remain constant as they rotate at the synchronous speed.

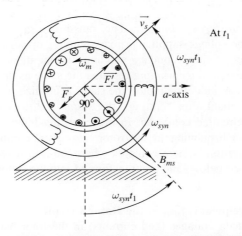

FIGURE 11.8 Rotor-produced mmf and the compensating mmf at time $t = t_1$.

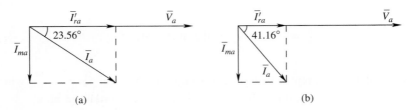

(a)   (b)

FIGURE 11.9   Example 11.2.

Example 11.2

Consider an induction machine that has 2 poles and is supplied by a rated voltage of 208 V (line-to-line, rms) at the frequency of 60 Hz. It is operating in steady state and is loaded to its rated torque. Neglect the stator leakage impedance and the rotor leakage flux. The per-phase magnetizing current is 4.0 A (rms). The current drawn per-phase is 10 A (rms) and is at an angle of 23.56 degrees (lagging). Calculate the per-phase current if the mechanical load decreases so that the slip speed is one-half that of the rated case.

Solution   We will consider the phase-*a* voltage to be the reference phasor. Therefore,

$$\overline{V}_a = \frac{208\sqrt{2}}{\sqrt{3}} \angle 0° = 169.8 \angle 0° \ V$$

It is given that at the rated load, as shown in Figure 11.9a, $\overline{I}_{ma} = 4.0\sqrt{2} \angle -90°$ A and $\overline{I}_a = 10.0\sqrt{2} \angle -23.56°$ A. From the phasor diagram in Figure 11.9a, $\overline{I}'_{ra} = 9.173\sqrt{2} \angle 0°$ A.

At one-half of the slip speed, the magnetizing current is the same but the amplitudes of the rotor-bar currents, and hence of $\overline{I}'_{ra}$ are reduced by one-half:

$$\overline{I}_{ma} = 4.0\sqrt{2} \angle -90° \ A \quad \text{and} \quad \overline{I}'_{ra} = 4.59\sqrt{2} \angle 0° A$$

Therefore, $\overline{I}_a = 6.1\sqrt{2} \angle -41.16°$ A, as shown in the phasor diagram of Figure 11.9b.

Revisiting the Transformer Analogy: The transformer equivalent circuit in Figure 11.4b illustrated that the voltage-supplied primary winding draws a compensating current to neutralize the effect of the secondary winding current in order to ensure that the resultant flux linking the *primary winding* remains the same as under the open-circuited condition. Similarly, in an induction motor, the stator neutralizes the rotor-produced field to ensure that the resultant flux "cutting" the *stator windings* remains the same as under the rotor open-circuited condition. In induction machines, this is how the stator keeps track of what is happening in the rotor. However, compared to transformers, induction machine operation is more complex where the rotor-cage quantities are at the slip frequency (discussed below) and are transformed into the stator-frequency quantities "seen" from the stator.

11.3.2.2   The Slip Frequency, f_{slip}, in the Rotor Circuit

The frequency of induced voltages (and currents) in the rotor bars can be obtained by considering Figure 11.10a. At $t = 0$, the bottom-most bar labeled "p" is being "cut" by

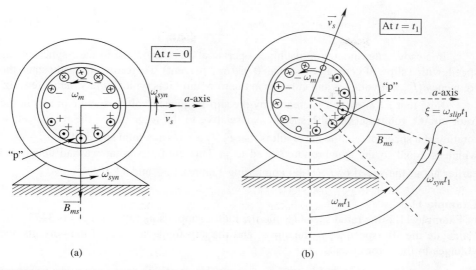

FIGURE 11.10 (a) Voltage induced in bar "p" at (a) $t = 0$; (b) $t = t_1$.

the positive peak flux density and has a positive induced voltage at the front-end. The $\vec{B}_{ms}(t)$ space vector, which is rotating with a speed of ω_{syn}, is "pulling ahead" at the slip speed ω_{slip} with respect to the rotor bar "p," which is rotating at ω_m. Therefore, as shown in Figure 11.10b, at sometime $t_1 > 0$, the angle between $\vec{B}_{ms}(t)$ and the rotor bar "p" is

$$\xi = \omega_{slip}t_1 \tag{11.22}$$

Therefore, the first time (call it $T_{2\pi}$) when the bar "p" is again being "cut" by the positive peak flux density is when $\zeta = 2\pi$. Therefore, from Equation 11.22,

$$T_{2\pi} = \frac{\xi(=2\pi)}{\omega_{slip}} \tag{11.23}$$

where $T_{2\pi}$ is the time-period between the two consecutive positive peaks of the induced voltage in the rotor bar "p." Therefore, the induced voltage in the rotor bar has a frequency (which we will call the slip frequency f_{slip}) which is the inverse of $T_{2\pi}$ in Equation 11.23:

$$f_{slip} = \frac{\omega_{slip}}{2\pi} \tag{11.24}$$

For convenience, we will define a unitless (dimensionless) quantity called slip, s, as the ratio of the slip speed to the synchronous speed:

$$s = \frac{\omega_{slip}}{\omega_{syn}} \tag{11.25}$$

Substituting for ω_{slip} from Equation 11.25 into Equation 11.24 and noting that $\omega_{syn} = 2\pi f$ (in a 2-pole machine),

$$f_{slip} = sf \tag{11.26}$$

In steady state, induction machines operate at ω_m, very close to their synchronous speed, with a slip s of generally less than 0.03 (or 3 percent). Therefore, in steady state, the frequency (f_{slip}) of voltages and currents in the rotor circuit is typically less than a few Hz.

Note that $\vec{F}_r(t)$, which is created by the slip-frequency voltages and currents in the rotor circuit, rotates at the slip speed ω_{slip}, relative to the rotor. Since the rotor itself is rotating at a speed of ω_m, the net result is that $\vec{F}_r(t)$ rotates at a total speed of $(\omega_{slip} + \omega_m)$, which is equal to the synchronous speed ω_{syn}. This confirms what we had concluded earlier about the speed of $\vec{F}_r(t)$ by comparing Figures 11.7 and 11.8.

Example 11.3
In Example 11.2, the rated speed (while the motor supplies its rated torque) is 3475 rpm. Calculate the slip speed ω_{slip}, the slip s, and the slip frequency f_{slip} of the currents and voltages in the rotor circuit.

Solution   This is a 2-pole motor. Therefore, at the rated frequency of 60 Hz, the rated synchronous speed, from Equation 11.8, is $\omega_{syn} = \omega = 2\pi \times 60 = 377$ rad/s. The rated speed is $\omega_{m,rated} = \frac{2\pi \times 3475}{60} = 363.9$ rad/s.

Therefore, $\omega_{slip,rated} = \omega_{syn,rated} - \omega_{m,rated} = 377.0 - 363.9 = 13.1$ rad/s.   From Equation 11.25.

$$slip\ s_{rated} = \frac{\omega_{slip,rated}}{\omega_{syn,rated}} = \frac{13.1}{377.0} = 0.0347 = 3.47\%$$

and, from Equation 11.26,

$$f_{slip,rated} = s_{rated}f = 2.08\ \text{Hz}.$$

11.3.2.3   Electromagnetic Torque
The electromagnetic torque on the rotor is produced by the interaction of the flux-density distribution represented by $\vec{B}_{ms}(t)$ in Figure 11.7a and the rotor-bar currents producing the mmf $\vec{F}_r(t)$. As in Chapter 10, it will be easier to calculate the torque produced on the rotor by first calculating the torque on the stator equivalent winding that produces the nullifying mmf $\vec{F}_r'(t)$. At $t=0$, this equivalent stator winding, sinusoidally distributed with N_s turns, has its axis along the $\vec{F}_r'(t)$ space vector, as shown in Figure 11.11. The winding also has a current $\hat{I}_r'$ flowing through it.

Following the derivation of the electromagnetic torque in Chapter 10, from Equation 10-5,

$$T_{em} = \pi r \ell \frac{N_s}{2} \hat{B}_{ms} \hat{I}_r' \tag{11.27}$$

The above equation can be written as

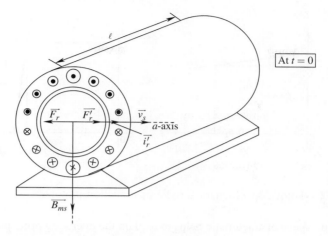

FIGURE 11.11 Calculation of electromagnetic torque.

$$T_{em} = k_t \, \hat{B}_{ms} \, \hat{I}_r' \quad \left(\text{where } k_t = \pi \, r\ell \frac{N_s}{2} \right) \tag{11.28}$$

where k_t is a constant, which depends on the machine design. The torque on the stator in Figure 11.11 acts in a clockwise direction, and the torque on the rotor is equal in magnitude and acts in a counter-clockwise direction.

The current peak $\hat{I}_r'$ depends linearly on the flux-density peak $\hat{B}_{ms}$ and the slip speed ω_{slip}, as expressed by Equation 11.21 ($\hat{I}_r' = k_i \, \hat{B}_{ms} \omega_{slip}$). Therefore, substituting for $\hat{I}_r'$ in Equation 11.28,

$$T_{em} = k_{t\omega} \, \hat{B}_{ms}^2 \omega_{slip} \quad (k_{t\omega} = k_t k_i) \tag{11.29}$$

where $k_{t\omega}$ is a machine torque constant. If the flux-density peak is maintained at its rated value in Equation 11.29,

$$T_{em} = k_{T\omega} \omega_{slip} \quad (k_{T\omega} = k_{t\omega} \, \hat{B}_{ms}^2) \tag{11.30}$$

where $k_{T\omega}$ is another torque constant of the machine.

Equation 11.30 expresses the torque-speed characteristic of induction machines. For a rated set of applied voltages, which result in $\omega_{syn,rated}$ and $\hat{B}_{ms,rated}$, the torque developed by the machine increases linearly with the slip speed ω_{slip} as the rotor slows down. This torque-speed characteristic is shown in Figure 11.12 in two different ways.

At zero torque, the slip speed ω_{slip} is zero, implying that the motor rotates at the synchronous speed. This is only a theoretical operating point because the motor's internal bearing friction and windage losses would require that a finite amount of electromagnetic torque be generated to overcome them. The torque-speed characteristic beyond the rated torque is shown dotted because the assumptions of neglecting stator leakage impedance and the rotor leakage inductance begin to break down.

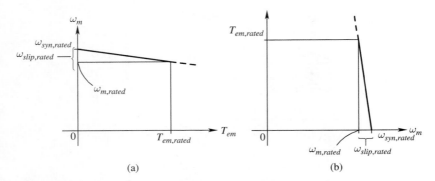

(a)   (b)

FIGURE 11.12   Torque speed characteristic of induction motors.

The torque-speed characteristic helps to explain the operating principle of induction machines, as illustrated in Figure 11.13. In steady state, the operating speed ω_{m1} is dictated by the intersection of the electromagnetic torque and the mechanical-load torque T_{L1}. If the load torque is increased to T_{L2}, the induction motor slows down to ω_{m2}, increasing the slip speed ω_{slip}. This increased slip speed results in higher induced voltages and currents in the rotor bars, and hence a higher electromagnetic torque is produced to meet the increase in mechanical load torque.

On a dynamic basis, the electromagnetic torque developed by the motor interacts with the shaft-coupled mechanical load, in accordance with the following mechanical-system equation:

$$\frac{d\omega_m}{dt} = \frac{T_{em} - T_L}{J_{eq}} \tag{11.31}$$

where J_{eq} is the combined motor-load inertia constant and T_L (generally a function of speed) is the torque of the mechanical load opposing the rotation. The acceleration torque is $(T_{em} - T_L)$.

Note that the electromagnetic torque developed by the motor equals the load torque in steady state. Often, the torque required to overcome friction and windage (including that of the motor itself) can be included lumped with the load torque.

Example 11.4

In Example 11.3, the rated torque supplied by the motor is 8 Nm. Calculate the torque constant $k_{T\omega}$, which linearly relates the torque developed by the motor to the slip speed.

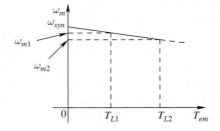

FIGURE 11.13   Operation of an induction motor.

Solution   From Equation 11.30, $k_{T\omega} = \frac{T_{em}}{\omega_{slip}}$. Therefore, using the rated conditions,

$$k_{T\omega} = \frac{T_{em,rated}}{\omega_{slip,rated}} = \frac{8.0}{13.1} = 0.61 \quad \frac{\text{Nm}}{\text{rad/s}}$$

The torque-speed characteristic is as shown in Figure 11.12, with the slope given above.

11.3.2.4   The Generator (Regenerative Braking) Mode of Operation

Induction machines can be used as generators, for example many wind-electric systems use induction generators to convert wind energy to electrical output that is fed into the utility grid. Most commonly, however, while slowing down, induction motors go into regenerative-braking mode (which, from the machine's standpoint, is the same as the generator mode), where the kinetic energy associated with the inertia of the mechanical system is converted into electrical output. In this mode of operation, the rotor speed exceeds the synchronous speed ($\omega_m > \omega_{syn}$) where both are in the same direction. Hence, $\omega_{slip} < 0$.

Under the condition of negative slip speed shown in Figure 11.14, the voltages and currents induced in the rotor bars are of opposite polarities and directions compared to those with positive slip speed in Figure 11.7a. Therefore, the electromagnetic torque on the rotor acts in a clockwise direction, opposing the rotation and thus slowing down the rotor. In this regenerative breaking mode, T_{em} in Equation 11.31 has a negative value.

Example 11.5

The induction machine of Example 11.2 is to produce the rated torque in the regenerative-braking mode. Draw the voltage and current phasors for phase-a.

Solution   With the assumption that the stator leakage impedance can be neglected, the magnetizing current is the same as in Example 11.2, $\bar{I}_{ma} = 4.0\sqrt{2} \angle -90°$ A as shown in Figure 11.15. However, since we are dealing with a regenerative-braking torque, $\bar{I}'_{ra} = -9.173\sqrt{2} \angle 0°$ A as shown in the phasor diagram of Figure 11.15. Hence, $\bar{I}_a = 10.0\sqrt{2} \angle -156.44°$ A as shown in Figure 11.15.

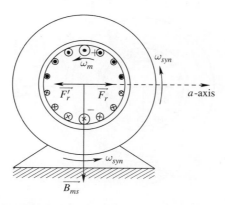

FIGURE 11.14   Regenerative braking in induction motors.

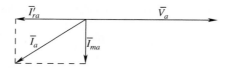

FIGURE 11.15   Example 11.5.

11.3.2.5   Reversing the Direction of Rotation

Reversing the sequence of applied voltages (*a-b-c* to *a-c-b*) causes the reversal of direction, as shown in Figure 11.16.

11.3.2.6   Including the Rotor Leakage Inductance

Up to the rated torque, the slip speed and the slip frequency in the rotor circuit are small, and hence it is reasonable to neglect the effect of the rotor leakage inductance. However, loading the machine beyond the rated torque results in larger slip speeds and slip frequencies, and the effect of the rotor leakage inductance should be included in the analysis, as described below.

Of all the flux produced by the currents in the rotor bars, a portion (which is called the leakage flux and is responsible for the rotor leakage inductance) does not completely cross the air gap and does not "cut" the stator windings. *First considering only the stator-established flux-density distribution* $\vec{B}_{ms}(t)$ at $t = 0$ as in Figure 11.6a, the top and the bottom bars are "cut" by the peak $\hat{B}_{ms}$ of the flux-density distribution, and due to this flux the voltages induced in them are the maximum. However (as shown in Figure 11.17a), the bar currents lag due to the inductive effect of the rotor leakage flux and are maximum in the bars that were "cut" by $\vec{B}_{ms}(t)$ sometime earlier. Therefore, the rotor mmf space vector $\vec{F}_r(t)$ in Figure 11.17a lags $\vec{B}_{ms}(t)$ by an angle $\frac{\pi}{2} + \theta_r$, where θ_r is called the rotor power factor angle.

At $t = 0$, the flux lines produced by the rotor currents in Figure 11.17b can be divided into two components: ϕ_{m, i_r}, which crosses the air gap and "cuts" the stator windings, and $\phi_{\ell r}$, the rotor leakage flux, which does *not* cross the air gap to "cut" the stator windings.

The stator excited by ideal voltage sources (and assuming that R_s and L_{ls} are zero) demands that the flux-density distribution $\vec{B}_{ms}(t)$ "cutting" it be unchanged. Therefore, additional stator currents, represented by $\vec{i}_r'(t)$ in Figure 11.17a, are drawn to produce $\phi_{m, i_r'}$ in Figure 11.17b to compensate for ϕ_{m, i_r} (but not to compensate for $\phi_{\ell r}$, whose

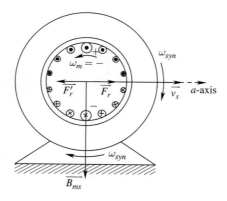

FIGURE 11.16   Reversing the direction of rotation in an induction motor.

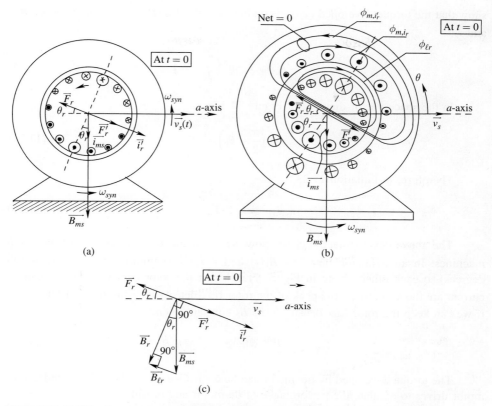

FIGURE 11.17 Space vectors with the effect of rotor leakage flux included.

existence the stator is unaware of), such that $\phi_{m,\,i'_r}$ is equal in magnitude and opposite in direction to $\phi_{m,\,i_r}$.

The additional currents drawn from the three stator phase windings can be represented by means of the equivalent stator winding with N_s turns and carrying a current $\hat{I}'_r$, as shown in Figure 11.17b. The resulting $\vec{F}_r$, $\vec{F}'_r$, and $\vec{i}'_r$ space vectors at $t = 0$ are shown in Figure 11.17c. The rotor bars are "cut" by the net flux-density distribution represented by $\vec{B}_r(t)$, shown in Figure 11.17c at $t = 0$, where

$$\vec{B}_r(t) = \vec{B}_{ms}(t) + \vec{B}_{\ell r}(t) \tag{11.32}$$

$\vec{B}_{\ell r}(t)$ represents in the air gap the rotor leakage flux density distribution (due to $\phi_{\ell r}$) which, for our purposes, is also assumed to be radial and sinusoidally distributed. Note that $\vec{B}_r$ is *not* created by the currents in the rotor bars; rather it is the flux-density distribution "cutting" the rotor bars.

The equivalent stator winding shown in Figure 11.17b has a current $\hat{I}'_r$ and is "cut" by the flux-density distribution represented by $\vec{B}_{ms}$. As shown in Figure 11.17c, the $\vec{B}_{ms}$ and $\vec{i}'_r$ space vectors are at an angle of $(\pi/2 - \theta_r)$ with respect to each other. Using a procedure similar to the one that led to the torque expression in Equation 11.28, we can show that the torque developed depends on the sine of the angle $(\pi/2 - \theta_r)$ between $\vec{B}_{ms}$ and $\vec{i}'_r$:

$$T_{em} = k_t \hat{B}_{ms} \hat{I}'_r \sin\left(\frac{\pi}{2} - \theta_r\right) \qquad (11.33)$$

In the space vector diagram of Figure 11.17c,

$$\hat{B}_{ms} \sin\left(\frac{\pi}{2} - \theta_r\right) = \hat{B}_r \qquad (11.34)$$

Therefore, in Equation 11.33,

$$T_{em} = k_t \hat{B}_r \hat{I}'_r \qquad (11.35)$$

The above development suggests how we can achieve vector control of induction machines. In an induction machine, $\vec{B}_r(t)$ and $\vec{i}'_r(t)$ are naturally at right angles (90 degrees) to each other. (Note in Figure 11.17b that the rotor bars with the maximum current are those "cutting" the peak of the rotor flux-density distribution $\hat{B}_r$.) Therefore, if we can keep the rotor flux-density peak $\hat{B}_r$ constant, then

$$T_{em} = k_T \hat{I}'_r \quad \text{where} \quad k_T = k_t \hat{B}_r \qquad (11.36)$$

The torque developed by the motor can be controlled by $\hat{I}'_r$. This allows induction-motor drives to emulate the performance of dc-motor and brushless-dc motor drives.

11.3.3 Per-Phase Steady-State Equivalent Circuit (Including Rotor Leakage)

The space vector diagram at $t = 0$ is shown in Figure 11.18a for the rated voltages applied. This results in the phasor diagram for phase-a in Figure 11.18b.

The current $\overline{I}'_{ra}$, which is lagging behind the applied voltage $\overline{V}_a$, can be represented as flowing through an inductive branch in the equivalent circuit of Figure 11.18c, where R_{eq} and L_{eq} are yet to be determined. For the above determination, assume that the rotor is blocked and that the voltages applied to the stator create the same conditions ($\vec{B}_{ms}$ with the same $\hat{B}_{ms}$ and at the same ω_{slip} with respect to the rotor) in the rotor circuit as in Figure 11.18. Therefore, in Figure 11.19a with the blocked rotor, we will apply stator voltages at the slip frequency $f_{slip}(= \omega_{slip}/2\pi)$ from Equation 11.8 and of amplitude $(\omega_{slip}/\omega_{syn})\hat{V}$ from Equation 11.7, as shown in Figures 11.19a and 11.19b.

The blocked-rotor bars, similar to those in the rotor turning at ω_m, are "cut" by an identical flux-density distribution (which has the same peak value $\hat{B}_{ms}$ and which rotates at the same slip speed ω_{slip} with respect to the rotor). The phasor diagram in the blocked-rotor case is shown in Figure 11.19b and the phase equivalent circuit is shown in Figure 11.19c. (The quantities at the stator terminals in the blocked-rotor case of Figure 11.19 are similar to those of a transformer primary, with its secondary winding short-circuited.) The current $\overline{I}'_{ra}$ in Figure 11.19c is at the slip frequency f_{slip} and flowing through an

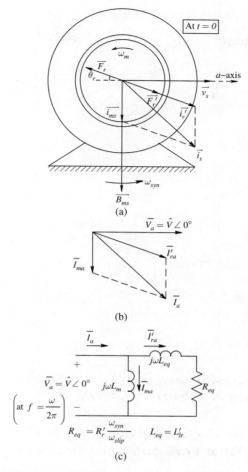

FIGURE 11.18 Rated voltage applied.

inductive branch, which consists of R'_r and $L'_{\ell r}$ connected in series. Note that R'_r and $L'_{\ell r}$ are the equivalent rotor resistance and the equivalent rotor leakage inductance, "seen" on a per-phase basis from the stator side. The impedance of the inductive branch with $\bar{I}'_{ra}$ in this blocked-rotor case is

$$Z_{eq,blocked} = R'_r + j\omega_{slip} L'_{\ell r} \tag{11.37}$$

The three-phase power loss in the bar resistances of the blocked rotor is

$$P_{r,loss} = 3R'_r (I'_{ra})^2 \tag{11.38}$$

where I'_{ra} is the rms value.

As far as the conditions "seen" by an observer sitting on the rotor are concerned, they are identical to the original case with the rotor turning at a speed ω_m but slipping at a speed ω_{slip} with respect to ω_{syn}. Therefore, in both cases, the current component $\bar{I}'_{ra}$ has the

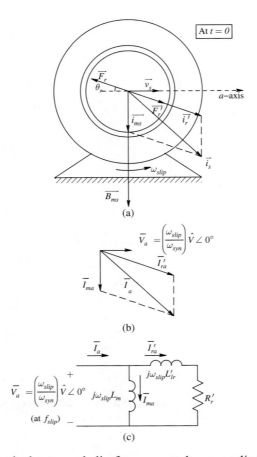

FIGURE 11.19   Blocked rotor and slip-frequency voltages applied.

same amplitude and the same phase angle with respect to the applied voltage. Therefore, in the original case of Figure 11.18, where the applied voltages are higher by a factor of $\frac{\omega_{syn}}{\omega_{slip}}$, the impedance must be higher by the same factor; that is, from Equation 11.37,

$$\underbrace{Z_{eq}}_{\text{at } f} = \frac{\omega_{syn}}{\omega_{slip}} \underbrace{(R'_r + j\omega_{slip}L'_{\ell r})}_{Z_{eq,\text{blocked at } f_{slip}}} = R'_r \frac{\omega_{syn}}{\omega_{slip}} + j\omega_{syn}L'_{\ell r} \tag{11.39}$$

Therefore, in the equivalent circuit of Figure 11.18c at frequency f, $R_{eq} = R'_r \frac{\omega_{syn}}{\omega_{slip}}$ and $L_{eq} = L'_{\ell r}$. The per-phase equivalent circuit of Figure 11.18c is repeated in Figure 11.20a, where $\omega_{syn} = \omega$ for a 2-pole machine. The power loss $P_{r,loss}$ in the rotor circuit in Figure 11.20a is the same as that given by Equation 11.38 for the blocked-rotor case of Figure 11.19c. Therefore, the resistance $R'_r \frac{\omega_{syn}}{\omega_{slip}}$ can be divided into two parts: R_r' and $R'_r \frac{\omega_m}{\omega_{slip}}$, as shown in Figure 11.20b, where $P_{r,loss}$ is lost as heat in R'_r and the power dissipation in $R'_r \frac{\omega_m}{\omega_{slip}}$, on a three-phase basis, gets converted into mechanical power (which also equals T_{em} times ω_m):

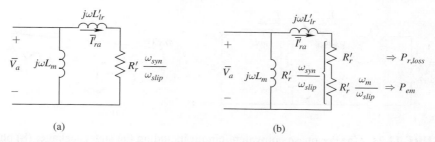

FIGURE 11.20 Splitting the rotor resistance into the loss component and power output component (neglecting the stator-winding leakage impedance).

$$P_{em} = 3 \frac{\omega_m}{\omega_{slip}} R'_r (I'_{ra})^2 = T_{em}\omega_m \tag{11.40}$$

Therefore,

$$T_{em} = 3R'_r \frac{(I'_{ra})^2}{\omega_{slip}} \tag{11.41}$$

From Equations 11.38 and 11.41,

$$\frac{P_{r,loss}}{T_{em}} = \omega_{slip} \tag{11.42}$$

This is an important relationship because it shows that to produce the desired torque T_{em} we should minimize the value of the slip speed in order to minimize the power loss in the rotor circuit.

Example 11.6
Consider a 60-Hz induction motor with $R'_r = 0.45 \ \Omega$ and $X'_{\ell r} = 0.85 \ \Omega$. The rated slip speed is 4 percent. Ignore the stator leakage impedance. Compare the torque at the rated slip speed by (a) ignoring the rotor leakage inductance and (b) including the rotor leakage inductance.

Solution   To calculate T_{em} at the rated slip speed we will make use of Equation 11.41, where I'_{ra} can be calculated from the per-phase equivalent circuit of Figure 11.20a. Ignoring the rotor leakage inductance,

$$I'_{ra}\big|_{L'_{\ell r}=0} = \frac{V_a}{R'_r \frac{\omega_{syn}}{\omega_{slip}}}, \quad \text{and from Equation 11.41} \quad T_{em}\big|_{L'_{\ell r}=0} = \frac{3R'_r}{\omega_{slip}} \frac{V_a^{\ 2}}{\left(R'_r \frac{\omega_{syn}}{\omega_{slip}}\right)^2}$$

Including the rotor leakage inductance,

$$I'_{ra} = \frac{V_a}{\sqrt{\left(R'_r \frac{\omega_{syn}}{\omega_{slip}}\right)^2 + (X'_{\ell r})^2}},$$

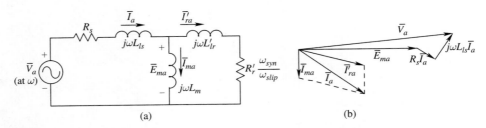

(a)   (b)

FIGURE 11.21   (a) Per phase equivalent circuit including the stator leakage; (b) phasor diagram.

and from Equation 11.41

$$T_{em} = \frac{3R'_r}{\omega_{slip}} \frac{V_a^2}{\left(R'_r \frac{\omega_{syn}}{\omega_{slip}}\right)^2 + \left(X'_{\ell r}\right)^2}$$

At the rated slip speed of 4%, $\frac{\omega_{slip}}{\omega_{syn}} = 0.04$. Therefore, comparing the above two expressions for torque by substituting the numerical values,

$$\frac{T_{em}\big|_{L'_{\ell r}=0}}{T_{em}} = \frac{\left(R'_r \frac{\omega_{syn}}{\omega_{slip}}\right)^2 + \left(X'_{\ell r}\right)^2}{\left(R'_r \frac{\omega_{syn}}{\omega_{slip}}\right)^2} = \frac{126.56^2 + 0.85^2}{126.56^2} \simeq 1.0$$

The above example shows that under normal operation when the motor is supplying a torque within its rated value, it does so at very small values of slip speed. Therefore, as shown in this example, we are justified in ignoring the effect of the rotor leakage inductance under normal operation. In high-performance applications requiring vector control, the effect of the rotor leakage inductance can be included, as discussed in Chapter 13.

11.3.3.7   Including the Stator Winding Resistance R_s and Leakage Inductance $L_{\ell s}$

Including the effect of the stator winding resistance R_s and the leakage inductance $L_{\ell s}$ is analogous to including the effect of primary winding impedance in the transformer equivalent circuit, as shown in Figure 11.21a.

In the per-phase equivalent circuit of Figure 11.21a, the applied voltage $\overline{V}_a$ is reduced by the voltage drop across the stator-winding leakage impedance to yield $\overline{E}_{ma}$:

$$\overline{E}_{ma} = \overline{V}_a - (R_s + j\omega L_{\ell s})\overline{I}_s \tag{11.43}$$

where $\overline{E}_{ma}$ represents the voltage induced in the stator phase-a by the rotating flux-density distribution $\vec{B}_{ms}(t)$. The phasor diagram with $\overline{E}_{ma}$ as the reference phasor is shown in Figure 11.21b.

11.4 TESTS TO OBTAIN THE PARAMETERS OF THE PER-PHASE EQUIVALENT CIRCUIT

The parameters of the per-phase equivalent circuit of Figure 11.21a are usually not supplied by the motor manufacturers. The three tests described below can be performed to estimate these parameters.

11.4.1 DC-Resistance Test to Estimate R_s

The stator resistance R_s can best be estimated by the dc measurement of the resistance between the two phases:

$$R_s(dc) = \frac{R_{phase-phase}}{2} \qquad (11.44)$$

This dc resistance value, measured by passing a dc current through two of the phases, can be modified by a skin-effect factor [1] to help estimate its line-frequency value more closely.

11.4.2 The No-Load Test to Estimate L_m

The magnetizing inductance L_m can be calculated from the no-load test. In this test, the motor is applied its rated stator voltages in steady state and no mechanical load is connected to the rotor shaft. Therefore, the rotor turns almost at the synchronous speed with $\omega_{slip} \cong 0$. Hence, the resistance $R'_r \frac{\omega_{syn}}{\omega_{slip}}$ in the equivalent circuit of Figure 11.21a becomes very large, allowing us to assume that $\overline{I}'_{ra} \cong 0$, as shown in Figure 11.22a.

The following quantities are measured: the per-phase rms voltage $V_a(= V_{LL}/\sqrt{3})$, the per-phase rms current I_a, and the three-phase power $P_{3-\phi}$ drawn by the motor. Subtracting the calculated power dissipation in R_s from the measured power, the remaining power $P_{FW,core}$ (the sum of the core losses, the stray losses, and the power to overcome friction and windage) is

$$P_{FW,core} = P_{3-\phi} - 3R_s I_a^2 \qquad (11.45)$$

With rated voltages applied to the motor, the above loss can be assumed to be a constant value that is independent of the motor loading.

Assuming that $L_m \gg L_{\ell s}$, the magnetizing inductance L_m can be calculated based on the per-phase reactive power Q from the following equation:

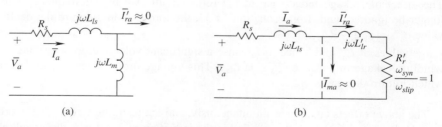

(a) (b)

FIGURE 11.22 (a) No-load test; (b) Blocked-rotor test.

$$Q = \sqrt{(V_a I_a)^2 - \left(\frac{P_{3-\phi}}{3}\right)^2} = (\omega L_m) I_a^2 \qquad (11.46)$$

11.4.3 Blocked-Rotor Test to Estimate R_r' and the Leakage Inductances

The blocked-rotor (or locked-rotor) test is conducted to determine both R_r', the rotor resistance "seen" from the stator on a per-phase basis, and the leakage inductances in the equivalent circuit of Figure 11.21a. Note that the rotor is blocked from turning and the stator is applied line-frequency, 3-phase voltages with a small magnitude such that the stator currents equal their rated value. With the rotor blocked, $\omega_m = 0$ and hence $\frac{\omega_{syn}}{\omega_{slip}} = 1$. The resulting equivalent impedance $\left(R_r' + j\omega L_{\ell r}'\right)$ in Figure 11.22b can be assumed to be much smaller than the magnetizing reactance $(j\omega L_m)$, which can be considered to be infinite. Therefore, by measuring V_a, I_a, and the three-phase power into the motor, we can calculate R_r' (having already estimated R_s previously) and $(L_{\ell s} + L_{\ell r}')$. In order to determine these two leakage inductances explicitly, we need to know their ratio, which depends on the design of the machine. As an approximation for general-purpose motors, we can assume that

$$L_{\ell s} \cong 23 L_{\ell r}' \qquad (11.47)$$

This allows both leakage inductances to be calculated explicitly.

11.5 INDUCTION MOTOR CHARACTERISTICS AT RATED VOLTAGES IN MAGNITUDE AND FREQUENCY

The typical torque-speed characteristic for general-purpose induction motors with name-plate (rated) values of applied voltages is shown in Figure 11.23a, where the normalized torque (as a ratio of its rated value) is plotted as a function of the rotor speed ω_m/ω_{syn}.

With no load connected to the shaft, the torque T_{em} demanded from the motor is very low (only enough to overcome the internal bearing friction and windage), and the rotor turns at a speed very close in value to the synchronous speed ω_{syn}. Up to the rated torque, the torque developed by the motor is linear with respect to ω_{slip}, a relationship given by Equation 11.30. Far beyond the rated condition, for which the machine is designed to operate in steady state, T_{em} no longer increases linearly with ω_{slip} for the following reasons:

1. The effect of leakage inductance in the rotor circuit at a higher frequency can no longer be ignored, and, from Equation 11.33, the torque is less due to the declining value of $\sin(\pi/2 - \theta_r)$.
2. Large values of I_{ra}' and hence of I_a cause a significant voltage drop across the stator winding leakage impedance $(R_s + j\omega L_{\ell s})$. This voltage drop causes E_{ma} to decrease, which in turn decreases $\hat{B}_{ms}$.

The above effects take place simultaneously, and the resulting torque characteristic for large values of ω_{slip} (which are avoided in the induction-motor drives discussed in the

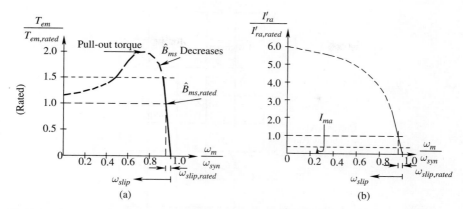

FIGURE 11.23 (a) Torque-speed characteristic; (b) Current-speed characteristic.

next chapter) is shown dotted in Figure 11.23a. The rated value of the slip-speed ω_{slip} at which the motor develops its rated torque is typically in a range of 0.03 to 0.05 times the synchronous speed ω_{syn}.

In the torque-speed characteristic of Figure 11.23a, the maximum torque that the motor can produce is called the pull-out (breakdown) torque. The torque when the rotor speed is zero is called the starting torque. The values of the pull-out and the starting torques, as a ratio of the rated torque, depend on the design class of the motor, as discussed in the next section.

Figure 11.23b shows the plot of the normalized rms current I'_{ra} as a function of the rotor speed. Up to the rated slip speed (up to the rated torque), I'_{ra} is linear with respect to the slip speed. This can be seen from Equation 11.21 (with $\hat{B}_{ms} = \hat{B}_{ms,rated}$):

$$\hat{I}'_r = (k_i \, \hat{B}_{ms,rated}) \, \omega_{slip} \tag{11.48}$$

Hence,

$$I'_{ra} = k_I \omega_{slip} \left(k_I = \frac{1}{\sqrt{2}} \frac{2}{3} k_i \, \hat{B}_{ms,rated} \right) \tag{11.49}$$

where k_I is a constant that linearly relates the slip speed to the rms current I'_{ra}. Notice that this plot is linear up to the rated slip speed, beyond which the effects of the stator and the rotor leakage inductances come into effect. At the rated operating point, the value of the rms magnetizing current I_{ma} is typically in a range of 20 to 40 percent of the per-phase stator rms current I_a. The magnetizing current I_{ma} remains relatively constant with speed, decreasing slightly at very large values of ω_{slip}. At or below the rated torque, the per-phase stator current magnitude I_a can be calculated by assuming the $\bar{I}'_{ra}$ and $\bar{I}_{ma}$ phasors to be at 90° with respect to each other; thus

$$I_a \cong \sqrt{I'^2_{ra} + I^2_{ma}} \quad \text{(below the rated torque)} \tag{11.50}$$

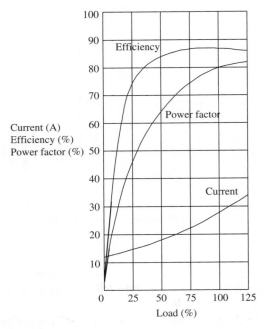

FIGURE 11.24   Typical performance curves for Design B 10 kW, 4-pole, 3-phase induction motor.

While delivering a torque higher than the rated torque, $\bar{I}'_{ra}$ is much larger in magnitude than the magnetizing current $\bar{I}_{ma}$ (also considering a large phase shift between the two). This allows the stator current to be approximated as follows:

$$I_a \cong I'_{ra} \quad \text{(above the rated torque)} \tag{11.51}$$

Figure 11.24 shows the typical variations of the power factor and the motor efficiency as a function of motor loading. These curves depend on the class and size of the motor and are discussed in Chapter 15, which deals with efficiency.

11.6   INDUCTION MOTORS OF NEMA DESIGN A, B, C, AND D

Three-phase induction machines are classified in the American Standards (NEMA) under five design letters: A, B, C, D, and F. Each design class of motors has different torque and current specifications. Figure 11.25 illustrates typical torque-speed curves for Design A, B, C and D motors; Design F motors have low pull-out and starting torques and thus are very limited in applications. As a ratio of the rated quantities, each design class specifies minimum values of pull-out and starting torques, and a maximum value of the starting current.

As noted previously, Design Class B motors are used most widely for general-purpose applications. These motors must have a minimum of a 200 percent pull-out torque.

Design A motors are similar to the general-purpose Design B motors except that they have a somewhat higher pull-out (breakdown) torque and a smaller full-load slip.

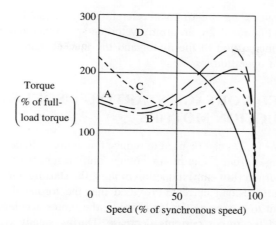

FIGURE 11.25 Typical torque-speed characteristics of NEMA Design A, B, C and D motors.

Design A motors are used when unusually low values of winding losses are required—in totally enclosed motors, for example.

Design C motors are high starting-torque, low starting-current machines. They also have a somewhat lower pull-out (breakdown) torque than Design A and B machines. Design C motors are almost always designed with double-cage rotor windings to enhance the rotor-winding skin effect.

Finally, Design D motors are high starting-torque, high-slip machines. The minimum starting torque is 275 percent of the rated torque. The starting torque in these motors can be assumed to be the same as the pull-out torque.

11.7 LINE START

It should be noted that the induction-motor drives, discussed in detail in the next chapter, are operated so as to keep ω_{slip} at low values. Hence, the dotted portions of the characteristics shown in Figure 11.23 are of no significance. However, if an induction motor is started from the line-voltage supply without an electronic power converter, it would at first draw 6 to 8 times its rated current, as shown in Figure 11.23b, limited mainly by the leakage inductances. Figure 11.26 shows that the available acceleration torque

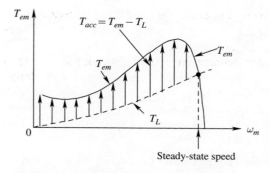

FIGURE 11.26 Available acceleration torque during start-up.

$T_{acc}(= T_{em} - T_L)$ causes the motor to accelerate from standstill, in accordance with Equation 11.31. In Figure 11.26, an arbitrary torque-speed characteristic of the load is assumed, and the intersection of the motor and the load characteristics determines the steady-state point of operation.

11.8   REDUCED VOLTAGE STARTING ("SOFT START") OF INDUCTION MOTORS

The circuit of Figure 11.27a can be used to reduce the motor voltages at starting, thereby reducing the starting currents. The motor voltage and current waveforms are shown in Figure 11.27b. In normal (low-slip) induction motors, the starting currents can be as large as 6 to 8 times the full-load current. Provided that the torque developed at reduced voltages is sufficient to overcome the load torque, the motor accelerates (the slip speed ω_{slip} decreases) and the motor currents decrease. During steady-state operation, each thyristor conducts for an entire half-cycle. Then, these thyristors can be shorted out (bypassed) by mechanical contactors, connected in parallel, to eliminate power losses in the thyristors due to a finite (1−2 V) conduction voltage drop across them.

11.9   ENERGY-SAVINGS IN LIGHTLY-LOADED MACHINES

The circuit of Figure 11.27a can also be used to minimize motor core losses in very lightly-loaded machines. Induction motors are designed such that it is most efficient to apply rated voltages at the full-load condition. With line-frequency voltages, the magnitude of the stator voltage at which the power loss is minimized slightly decreases with a decreasing load. Therefore, it is possible to use the circuit of Figure 11.27a to reduce the applied voltages at reduced loads and hence save energy. The amount of energy saved is significant (compared to extra losses in the motor due to current harmonics and in the thyristors due to a finite conduction voltage drop) only if the motor operates at very light loads for substantial periods of time. In applications where reduced voltage starting ("soft start") is required, the power switches are already implemented and only the controller for

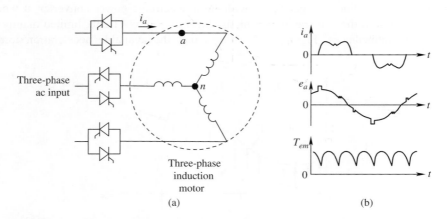

FIGURE 11.27   Stator voltage control (a) circuit; (b) waveforms.

the minimum power loss needs to be added. In such cases, the concept of reducing the voltage may be economical.

11.10 DOUBLY-FED INDUCTION GENERATORS (DFIG) IN WIND TURBINES

In utility-scale wind-turbines, common configurations are to use squirrel-cage induction or PMAC generators, shown by a block-diagram in Figure 11.28a and in more detail in Figure 11.28b for wind-turbine applications [4].

   The advantage of this configuration is that there is no need for mechanical contacts, that is, slip-rings and brushes, which are needed for the configuration described in this section. Moreover, there is total flexibility of the speed of turbine rotation, which is decoupled by the power-electronics interface from the synchronous speed dictated by the grid frequency. In this arrangement, the power electronic interface can also supply (or draw) reactive power to the grid for voltage stability. However, on the negative side, the entire power flows through the power-electronics interface that is still expensive but is declining in its relative cost.

   Another common configuration uses induction generators with wound rotors, as shown in Figure 11.29a in a block-diagram form in Figure 11.29b in a greater detail for wind-turbine applications [4]. The stator of such generators is directly connected to the

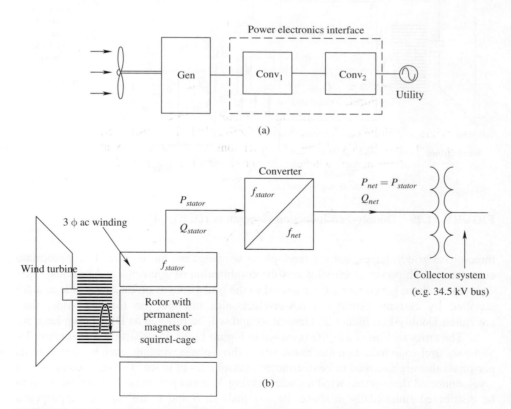

FIGURE 11.28   Wind-turbines with a complete power-electronics interface [4].

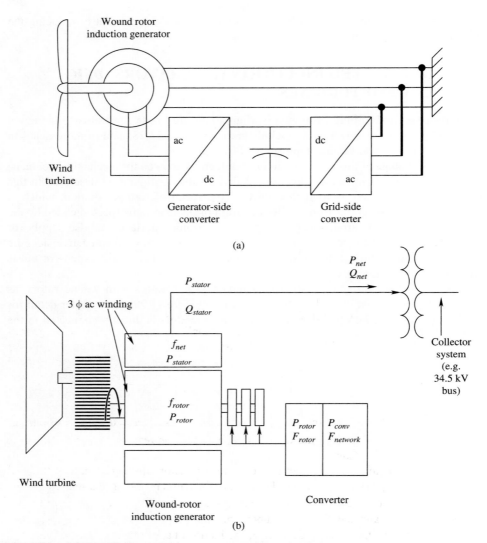

FIGURE 11.29   Doubly-Fed Induction Generators (DFIG) [4].

three-phase grid voltages, but the three-phase windings on the rotor are fed appropriate currents through power electronics and the combination of slip-rings and brushes.

Since these generators are connected to the grid-voltages on the stator side and are supplied by currents through a power-electronics interface on the rotor-side, they are called Doubly-Fed Induction Generators and will be referred as DFIG from here on.

The cross-section of a DFIG is shown in Figure 11.30. It consists of a stator, similar to the squirrel-cage induction machines, with a three-phase winding, each having N_s turns per-phase that are assumed to be distributed sinusoidally in space. The rotor consists of a wye-connected three-phase winding, each having N_r turns per-phase that are assumed to be distributed sinusoidally in space. Its terminals A, B and C are supplied appropriate currents through slip-rings and brushes, as shown in Figure 11.29b.

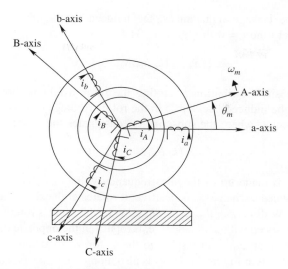

FIGURE 11.30 Cross-section of DFIGURE.

For our analysis, we will assume that this DFIG is operating under a balanced sinusoidal steady-state condition, with its stator supplied by the 60-Hz grid frequency voltages. In this simplified analysis, we will assume a 2-pole machine and neglect the stator resistance R_s and the leakage inductance $L_{\ell s}$. In this analysis, we will assume the motoring convention, where the currents are defined to be entering the stator- and the rotor-winding terminals, and the electromagnetic torque delivered to the shaft of the machine is defined to be positive.

Let us also assume that the voltage in phase-A peaks at $\omega t = 0$. At this instant, as shown in Figure 11.31, $\vec{v}_s$ is along the phase-A axis and the resulting $\vec{i}_{ms}$ and $\vec{B}_{ms}$ space vectors are vertical as shown.

All space vectors, with respect to the stationary stator windings, rotate at the synchronous speed ω_{syn} in CCW direction. The rotor of the DFIG is turning at a speed ω_m in the CCW direction, where the slip speed $\omega_{slip}(= \omega_{syn} - \omega_m)$ is positive in the sub-synchronous ($\omega_m < \omega_{syn}$) mode, and negative in the super-synchronous ($\omega_m > \omega_{syn}$) mode.

Based on Equation 9.41 in Chapter 9, the induced back-emf in the stator windings can be represented by the following space vector (neglecting the stator winding

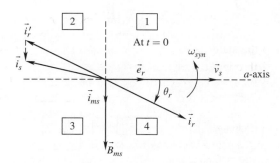

FIGURE 11.31 DFIG space vectors at time $t = 0$; drawn with $\omega_{slip} = +$.

resistances and the leakage inductances, the induced back-emfs are the same as the applied voltages) at time $t = 0$ in Figure 11.31:

$$\vec{e}_s = \vec{v}_s = \left(k_e \hat{B}_{ms} N_s \omega_{syn}\right) \angle 0° \tag{11.52}$$

The same flux-density distribution is cutting the rotor windings, however, at the slip speed. Therefore, the induced back-emfs in the rotor windings can be represented by the following space vector, where the subscript "r" designates the rotor:

$$\vec{e}_r = \left(k_e \hat{B}_{ms} N_r \omega_{slip}\right) \angle 0° \tag{11.53}$$

Note that $\vec{e}_r$ is made up of the slip-frequency voltages $e_A(t)$, $e_B(t)$, and $e_C(t)$. At sub-synchronous speeds, when ω_{slip} is positive, it rotates at the slip-speed ω_{slip} relative to the rotor in the CCW direction, the same direction as the rotor is rotating in; otherwise at super-synchronous speed with ω_{slip} negative, it rotates in the opposite direction. Since the rotor itself is rotating at ω_m, with respect to the stator, $\vec{e}_r$ rotates at $\omega_{syn}(= \omega_m + \omega_{slip})$, similar to $\vec{v}_s$. At $\omega t = 0$ in Figure 11.31, $\vec{e}_r$ is also along the same axis as $\vec{e}_s(= \vec{v}_s)$ if ω_{slip} is positive (otherwise, opposite), regardless of where the rotor A-axis may be in Figure 11.30 (why? see the homework problem).

We should note that when $\vec{e}_r$ rotates in the CCW direction (at sub-synchronous speeds), same as $\vec{v}_s$, and the phase sequence of the slip-frequency voltages induced in the rotor windings is A-B-C, same as the a-b-c sequence applied to the stator windings. However, at super-synchronous speeds, $\vec{e}_r$ rotates in the CW direction, opposite to $\vec{v}_s$, and the phase sequence of the slip-frequency voltages induced in the rotor windings is A-C-B, negative of the a-b-c sequence applied to the stator windings.

Appropriate slip-frequency voltages $\vec{v}_r$ are applied from the power electronic converter, through slip rings/brushes, as shown in the one-line diagram of Figure 11.32, in order to control the current $\vec{i}_r$ to be as desired in Figure 11.31.

Assuming the current direction to be going into the rotor windings as shown,

$$\vec{i}_r = \frac{(\vec{v}_r - \vec{e}_r)}{(R_r + j\omega_{slip} L_{\ell r})} \quad \text{at slip-frequency} \tag{11.54}$$

In Figure 11.31, to nullify the mmf produced by the rotor currents, additional currents drawn from the stator result in

$$\vec{i}_r' = -\left(\frac{N_r}{N_s}\right)\vec{i}_r \tag{11.55}$$

The current into the stator is

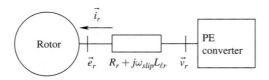

FIGURE 11.32   Rotor circuit one-line diagram at slip-frequency.

$$\vec{i}_s = \vec{i}_{ms} + \vec{i}'_r \tag{11.56}$$

Based on the space vectors shown in Figure 11.31, the complex power $S_s(= P_s + jQ_s)$ *into* the stator is

$$S_s = P_s + jQ_s = \frac{2}{3}\vec{v}_s\vec{i}_s = \frac{2}{3}\vec{v}_s(\vec{i}_{ms} + \vec{i}'_r) \tag{11.57}$$

From Figure 11.31,

$$\vec{i}_r = \hat{I}_r \angle - \theta_r \; (\theta_r \text{ has a positive value as shown in Figure 11.31}) \tag{11.58}$$

Using Equations 11.55 and 11.58 into Equation 11.57, and recognizing that in Figure 11.31 $\vec{v}_s = \hat{V}_s \angle 0°$, the real power into the stator is

$$P_s = \frac{2}{3}\hat{V}_s \text{Re}[\vec{i}'_r] = -\frac{2}{3}\hat{V}_s\hat{I}_r\frac{N_r}{N_s}\cos\theta_r \tag{11.59}$$

The reactive power into the stator is due to the magnetizing current $\vec{i}_{ms}$ and $\vec{i}'_r$. Therefore, we can write the reactive power Q_s *into* the stator as

$$Q_s = Q_{mag} + Q'_r \tag{11.60}$$

where,

$$Q_{mag} = \frac{2}{3}\hat{V}_s\hat{I}_{ms} \tag{11.61}$$

and,

$$Q'_r = \frac{2}{3}\hat{V}_s\text{Im}[\vec{i}_r'] = -\frac{2}{3}\hat{V}_s\hat{I}_r\frac{N_r}{N_s}\sin\theta_r \tag{11.62}$$

Similarly, the complex power $S_r(= P_r + jQ_r)$ *into* the rotor back-emfs is

$$S_r = P_r + jQ_r = \frac{2}{3}\vec{e}_r\vec{i}^*_r \tag{11.63}$$

where, recognizing that in Figure 11.31 $\vec{e}_r = \hat{E}_r\angle 0°$,

$$P_r = \frac{2}{3}\hat{E}_r\text{Re}[\vec{i}^*_r] = \frac{2}{3}\hat{E}_r\hat{I}_r\cos\theta_r \tag{11.64}$$

and,

$$Q_r = \frac{2}{3}\hat{E}_r\text{Im}[\vec{i}^*_r] = \frac{2}{3}\hat{E}_r\hat{I}_r\sin\theta_r \tag{11.65}$$

From Equations 11.59 and 11.64, and making use of Equations 11.52 and 11.53,

$$\frac{P_s}{P_r} = -\frac{\omega_{syn}}{\omega_{slip}} = -\frac{1}{s} \tag{11.66}$$

The total real electrical power into this doubly-fed induction machine, which gets converted into the output mechanical power to the shaft, is

$$P_{em} = P_s + P_r = P_s(1 - s) \tag{11.67}$$

Comparing the reactive powers,

$$\frac{Q_r'}{Q_r} = -\frac{\omega_{syn}}{\omega_{slip}} = -\frac{1}{s} \tag{11.68}$$

which shows that the reactive power input of Q_r into the rotor back-emfs is amplified by a factor of $(1/s)$ at the stator in magnitude. Therefore, from Equation 11.60,

$$Q_s = Q_{mag} + Q_r' = Q_{mag} - \frac{Q_r}{s} \tag{11.69}$$

Figure 11.33 shows the flow of real and reactive powers, where the real and reactive power losses associated with the resistances and the leakage inductances in the stator and the rotor circuits are not included, and a motoring convention is used to define the flows. It should be noted that Q_r is unrelated to the reactive power associated with the grid-side converter shown in Figure 11.33.

Table 11.1 shows various operating conditions in the sub-synchronous (sub-syn) and the super-synchronous (super-syn) modes.

In the analysis above, real and reactive powers associated with the stator and the rotor resistances and leakage inductances must be added for a complete analysis (see homework problems).

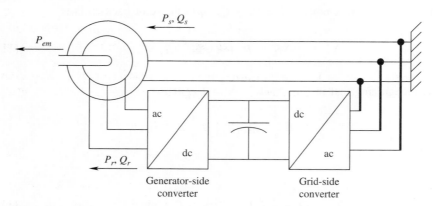

FIGURE 11.33 Flows of real and reactive powers in a DFIG using a motoring convention; the real and reactive power losses associated with the resistances and the leakage inductances in the stator and the rotor circuits are not included.

TABLE 11.1 Various Operating Modes of DFIGURE (note that the Row number 1 corresponds to the space vectors in Figure 11.31)

Row	ω_{slip},s (Speed)	$\vec{e}_r$ (at $t = 0$)	$\vec{i}_r$ in Quadrant	P_s (mode)	P_r	Q_r	Q'_r
1	+ (sub-syn)	+	4	$P_s = -$ (generating)	+	+	−
2	+ (sub-syn)	+	1	$P_s = -$ (generating)	+	−	+
3	+ (sub-syn)	+	3	$P_s = +$ (motoring)	−	+	−
4	+ (sub-syn)	+	2	$P_s = +$ (motoring)	−	−	+
5	− (super-syn)	−	4	$P_s = -$ (generating)	−	−	−
6	− (super-syn)	−	1	$P_s = -$ (generating)	−	+	+
7	− (super-syn)	−	3	$P_s = +$ (motoring)	+	−	−
8	− (super-syn)	−	2	$P_s = +$ (motoring)	+	+	+

SUMMARY/REVIEW QUESTIONS

1. Describe the construction of squirrel-cage induction machines.
2. With the rated voltages applied, what does the magnetizing current depend on? Does this current, to a significant extent, depend on the mechanical load on the motor? How large is it in relation to the rated motor current?
3. Draw the space vector diagram at $t = 0$, and the corresponding phasor diagram, assuming the rotor to be open-circuited.
4. Under a balanced, three-phase, sinusoidal steady-state excitation, what is the speed of the rotating flux-density distribution called? How is this speed related to the angular frequency of the electrical excitation in a p-pole machine?
5. In our analysis, why did we initially assume the stator leakage impedance to be zero? How does the transformer analogy, with the primary winding leakage impedance assumed to be zero, help? Under the assumption that the stator leakage impedance is zero, is $\vec{B}_{ms}(t)$ completely independent of the motor loading?
6. What is the definition of the slip speed ω_{slip}? Does ω_{slip} depend on the number of poles? How large is the rated slip speed, compared to the rated synchronous speed?
7. Write the expressions for the voltage and the current (assuming the rotor leakage inductance to be zero) in a rotor bar located at an angle θ from the peak of $\vec{B}_{ms}$.
8. The rotor bars located around the periphery of the rotor are of uniform cross-section. In spite of this, what allows us to represent the mmf produced by the rotor bar currents by a space vector $\vec{F}_r(t)$ at any time t?
9. Assuming the stator leakage impedance and the rotor inductance to be zero, draw the space vector diagram, the phasor diagram, and the per-phase equivalent circuit of a loaded induction motor.
10. In the equivalent circuit of Problem 9, what quantities does the rotor-bar current peak, represented by $\hat{I}'_{ra}$, depend on?
11. What is the frequency of voltages and currents in the rotor circuit called? How is it related to the slip speed? Does it depend on the number of poles?
12. What is definition of slip s, and how does it relate the frequency of voltages and currents in the stator circuit to that in the rotor circuit?
13. What is the speed of rotation of the mmf distribution produced by the rotor bar currents: (a) with respect to the rotor? (b) in the air gap with respect to a stationary observer?

14. Assuming $L'_{\ell r}$ to be zero, what is the expression for the torque T_{em} produced? How and why does it depend on ω_{slip} and $\hat{B}_{ms}$? Draw the torque-speed characteristic.

15. Assuming $L'_{\ell r}$ to be zero, explain how induction motors meet load-torque demand.

16. What makes an induction machine go into the regenerative-braking mode? Draw the space vectors and the corresponding phasors under the regenerative-braking condition.

17. Can an induction machine be operated as a generator that feeds into a passive load, for example a bank of three-phase resistors?

18. How is it possible to reverse the direction of rotation of an induction machine?

19. Explain the effect of including the rotor leakage flux by means of a space vector diagram.

20. How do we derive the torque expression, including the effect of $L'_{\ell r}$?

21. What is $\vec{B}_r(t)$, and how does it differ from $\vec{B}_{ms}(t)$? Is $\vec{B}_r(t)$ perpendicular to the $\vec{F}_r(t)$ space vector?

22. Including the rotor leakage flux, which rotor bars have the highest currents at any instant of time?

23. What clue do we have for the vector control of induction machines, to emulate the performance of brush-type and brush-less dc motors discussed in Chapters 7 and 10?

24. Describe how to obtain the per-phase equivalent circuit, including the effect of the rotor leakage flux.

25. What is the difference between $\overline{I}'_{ra}$ in Figure 11.18c and in Figure 11.19c, in terms of its frequency, magnitude, and phase angle?

26. Is the torque expression in Equation 11.41 valid in the presence of the rotor leakage inductance and the stator leakage impedance?

27. When producing a desired torque T_{em}, what is the power loss in the rotor circuit proportional to?

28. Draw the per-phase equivalent circuit, including the stator leakage impedance.

29. Describe the tests and the procedure to obtain the parameters of the per-phase equivalent circuit.

30. In steady state, how is the mechanical torque at the shaft different from the electromechanical torque T_{em} developed by the machine?

31. Do induction machines have voltage and torque constants similar to other machines that we have studied so far? If so, write their expressions.

32. Plot the torque-speed characteristic of an induction motor for applied rated voltages. Describe various portions of this characteristic.

33. What are the various classes of induction machines? Briefly describe their differences.

34. What are the problems associated with the line-starting of induction motors? Why is the starting current so high?

35. Why is reduced-voltage starting used? Show the circuit implementation, and discuss the pros and cons of using it to save energy.

REFERENCES

1. N. Mohan, T. Undeland, and W. Robbins, *Power Electronics: Converters, Applications, and Design*, 2nd ed. (New York: John Wiley & Sons, 1995).

2. A. E. Fitzgerald, Charles Kingsley, and Stephen Umans, *Electric Machinery*, 5th ed. (New York: McGraw Hill, 1990).

3. G. R. Slemon, *Electric Machines and Drives* (Addison-Wesley, Inc., 1992).

4. Kara Clark, Nicholas W. Miller, and Juan J. Sanchez-Gasca, *Modeling of GE Wind Turbine-Generators for Grid Studies*, GE Energy Report, Version 4.4, September 9, 2009.

PROBLEMS

11.1 Consider a three-phase, 2-pole induction machine. Neglect the stator winding resistance and the leakage inductance. The rated voltage is 208 V (line-line, rms) at 60 Hz. $L_m = 60$ mH, and the peak flux density in the air gap is 0.85 T. Consider that the phase-a voltage reaches its positive peak at $\omega t = 0$. Assuming that the rotor circuit is somehow open-circuited, calculate and draw the following space vectors at $\omega t = 0$ and at $\omega t = 60°$: $\vec{v}_s$, $\vec{i}_{ms}$, and $\vec{B}_{ms}$. Draw the phasor diagram with $\overline{V}_a$ and $\overline{I}_{ma}$. What is the relationship between $\hat{B}_{ms}$, $\hat{I}_{ms}$, and $\hat{I}_m$?

11.2 Calculate the synchronous speed in machines with a rated frequency of 60 Hz and with the following number of poles p: 2, 4, 6, 8, and 12.

11.3 The machines in Problem 11.2 produce the rated torque at a slip $s = 4$ percent, when supplied with rated voltages. Under the rated torque condition, calculate in each case the slip speed ω_{slip} in rad/s and the frequency f_{slip} (in Hz) of the currents and voltages in the rotor circuit.

11.4 In the transformer of Figure 11.4a, each air gap has a length $\ell_g = 1.0$ mm. The core iron can be assumed to have an infinite permeability. $N_1 = 100$ turns and $N_2 = 50$ turns. In the air gap, $\hat{B}_g = 1.1$ T and $v_1(t) = 100\sqrt{2} \cos \omega t$ at a frequency of 60 Hz. The leakage impedance of the primary winding can be neglected. With the secondary winding open-circuited, calculate and plot $i_m(t)$, $\phi_m(t)$, and the induced voltage $e_2(t)$ in the secondary winding due to $\phi_m(t)$ and $v_1(t)$.

11.5 In Example 11.1, calculate the magnetizing inductance L_m.

11.6 In an induction machine, the torque constant $k_{T\omega}$ (in Equation 11.30) and the rotor resistance R'_r are specified. Calculate $\hat{I}'_r$ as a function of ω_{slip}, in terms of $k_{T\omega}$ and R'_r, for torques below the rated value. Assume that the flux-density in the air gap is at its rated value. Hint: use Equation 11.41.

11.7 An induction motor produces rated torque at a slip speed of 100 rpm. If a new machine is built with bars of a material that has twice the resistivity of the old machine (and nothing else is changed), calculate the slip speed in the new machine when it is loaded to the rated torque.

11.8 In the transformer circuit of Figure 11.4b, the load on the secondary winding is a pure resistance R_L. Show that the emf induced in the secondary winding (due to the time-derivative of the combination of ϕ_m and the secondary-winding leakage flux) is in phase with the secondary current i_2. Note: this is analogous to the induction-motor case, where the rotor leakage flux is included and the current is maximum in the bar, which is "cut" by $\hat{B}_r$, the peak of the rotor flux-density distribution (represented by $\vec{B}_r$).

11.9 In a 60-Hz, 208 V (line-line, rms), 5-kW motor, $R'_r = 0.45$ Ω and $X'_{\ell r} = 0.83$ Ω. The rated torque is developed at the slip $s = 0.04$. Assuming that the motor is supplied with rated voltages and is delivering the rated torque, calculate the rotor power factor angle. What is $\hat{B}_r / \hat{B}_{ms}$?

11.10 In a 2-pole, 208 V (line-to-line, rms), 60-Hz, motor, $R_s = 0.5$ Ω, $R'_r = 0.45$ Ω, $X_{\ell s} = 0.6$ Ω, and $X'_{\ell r} = 0.83$ Ω. The magnetizing reactance $X_m = 28.5$ Ω. This motor is supplied by its rated voltages. The rated torque is developed at the slip

$s = 0.04$. At the rated torque, calculate the rotor power loss, the input current, and the input power factor of operation.

11.11 In a 208-V (line-to-line, rms), 60-Hz, 5-kW motor, tests are carried out with the following results: $R_{phase-phase} = 1.1 \ \Omega$. No-Load Test: applied voltages of 208 V (line-line, rms), $I_a = 6.5 \ A$, and $P_{no-load,3-phase} = 175 \ W$. Blocked-Rotor Test: applied voltages of 53 V (line-line, rms), $I_a = 18.2 \ A$, and $P_{blocked,3-phase} = 900 \ W$. Estimate the per-phase equivalent circuit parameters.

11.12 In Figure 11.31 of DFIG, explain why $\vec{e}_r$ is in phase with (or 180 opposite to) $\vec{v}_s$.

11.13 Draw the appropriate space vectors and the phasors corresponding to the generator-mode of operation in a sub-synchronous mode to confirm the entries in Table 11.1.

11.14 Draw the appropriate space vectors and the phasors corresponding to the generator-mode of operation in a super-synchronous mode to confirm the entries in Table 11.1.

11.15 A 6-pole, 3-phase DFIG is rated at $V_{LL}(\text{rms}) = 480$ V at 60 Hz. In the per-phase equivalent circuit of Figure 11.21, its parameters are as follows: $R_s = 0.008 \ \Omega$, $X_{\ell s} = 0.1 \ \Omega$, $X_m = 2.3 \ \Omega$, $R_r' = 0.125 \ \Omega$, and $X_{\ell r}' = 0.1 \ 5 \ \Omega$. Assume the friction, windage, and iron losses to be negligible. The rotor to the stator winding ratio is 2.5 to 1.0. This generator is supplying real power of 100 kW and the reactive power of 50 kV to the grid at a rotational speed of 1,320 rpm. Calculate the voltages that should be applied to the rotor circuit from the power-electronics converter. Calculate the power and the reactive power losses within the machine.

12

INDUCTION-MOTOR DRIVES: SPEED CONTROL

12.1 INTRODUCTION

Induction-motor drives are used in the process-control industry to adjust the speeds of fans, compressors, pumps, and the like. In many applications, the capability to vary speed efficiently can lead to large savings in energy. Adjustable-speed induction-motor drives are also used for electric traction and for motion control to automate factories.

Figure 12.1 shows the block diagram of an adjustable-speed induction-motor drive. The utility input can be either single-phase or three-phase. It is converted by the power-processing unit into three-phase voltages of appropriate magnitude and frequency, based on the controller input. In most general-purpose adjustable-speed drives (ASDs), the speed is not sensed, and hence the speed-sensor block and its input to the controller are shown dotted.

It is possible to adjust the induction-motor speed by controlling only the magnitude of the line-frequency voltages applied to the motor. For this purpose, a thyristor circuit, similar to that for "soft-start" in Figure 11.27a, can be used. Although simple and inexpensive to implement, this method is extremely energy-inefficient if the speed is to be varied over a wide range. Also, there are various other methods of speed control, but they require wound-rotor induction motors. Their description can be found in references listed at the end of Chapter 11. Our focus in this chapter is on examining energy-efficient speed control of squirrel-cage induction motors over a wide range. The emphasis is on general-purpose speed control rather than precise control of position using vector control, which is discussed in Chapter 13.

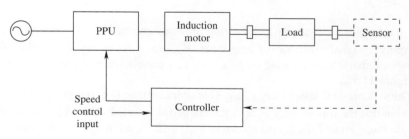

FIGURE 12.1 Block diagram of an induction-motor drive.

12.2   CONDITIONS FOR EFFICIENT SPEED CONTROL OVER A WIDE RANGE

In the block diagram of the induction-motor drive shown in Figure 12.1, we find that an energy-efficient system requires that both the power-processing unit and the induction motor maintain high energy efficiency over a wide range of speed and torque conditions. In Chapter 4, it was shown that the switch-mode techniques result in very high efficiencies of the power-processing units. Therefore, the focus in this section will be on achieving high efficiency of induction motors over a wide range of speed and torque.

We will begin this discussion by first considering the case in which an induction motor is applied the rated voltages (line-frequency sinusoidal voltages of the rated amplitude $\hat{V}_{rated}$ and the rated frequency f_{rated}, which are the same as the name-plate values). In Chapter 11, we derived the following expressions for a line-fed induction motor:

$$\frac{P_{r,loss}}{T_{em}} = \omega_{slip} \quad \text{(Equation 11.42, repeated)} \tag{12.1}$$

and

$$T_{em} = k_{t\omega}\,\hat{B}_{ms}^2\,\omega_{slip} \quad \text{(Equation 11.29, repeated)} \tag{12.2}$$

Equation 12.1 shows that to meet the load-torque demand ($T_{em} = T_L$), the motor should be operated with as small a slip speed ω_{slip} as possible in order to minimize power loss in the rotor circuit (this also minimizes the loss in the stator resistance). Equation 12.2 can be written as

$$\omega_{slip} = \frac{T_{em}}{k_{t\omega}\,\hat{B}_{ms}^2} \tag{12.3}$$

This shows that to minimize ω_{slip} at the required torque, the peak flux density $\hat{B}_{ms}$ should be kept as high as possible, the highest value being $\hat{B}_{ms,rated}$, for which the motor is designed and beyond which the iron in the motor will become saturated. (For additional discussion, please see Section 12.9.) Therefore, keeping $\hat{B}_{ms}$ constant at its rated value, the electromagnetic torque developed by the motor depends linearly on the slip speed ω_{slip}:

$$T_{em} = k_{T\omega}\,\omega_{slip} \quad \left(k_{T\omega} = k_{t\omega}\,\hat{B}_{ms,rated}^2\right) \tag{12.4}$$

This is the similar to Equation 11.30 of the previous chapter.

Applying rated voltages (of amplitude $\hat{V}_{rated}$ and frequency f_{rated}), the resulting torque-speed characteristic based on Equation 12.4 are shown in Figure 12.2a, repeated from Figure 11.12a.

The synchronous speed is $\omega_{syn,rated}$. This characteristic is a straight line based on the assumption that the flux-density peak is maintained at its rated value $\hat{B}_{ms,\,rated}$ throughout the torque range up to $T_{em,rated}$. As shown in Figure 12.2a, a family of such characteristics corresponding to various frequencies $f_3 < f_2 < f_1 < f_{rated}$ can be achieved (assuming that the flux-density peak is maintained throughout at its rated value $\hat{B}_{ms,\,rated}$, as

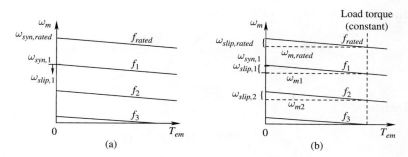

FIGURE 12.2 Operation characteristics with constant $\hat{B}_{ms} = \hat{B}_{ms,rated}$.

discussed in the next section). Focusing on the frequency f_1 corresponding to one of the characteristics in Figure 12.2a, the synchronous speed at which the flux-density distribution in the air gap rotates is given by

$$\omega_{syn,1} = \frac{2\pi f_1}{p/2} \tag{12.5}$$

Therefore, at a rotor speed $\omega_m(<\omega_{syn,1})$, the slip speed, measured with respect to the synchronous speed $\omega_{syn,1}$, is

$$\omega_{slip,1} = \omega_{syn,1} - \omega_m \tag{12.6}$$

Using the above $\omega_{slip,1}$ in Equation 12.4, the torque-speed characteristic at f_1 has the same slope as at f_{rated}. This shows that the characteristics at various frequencies are parallel to each other, as shown in Figure 12.2a. Considering a load whose torque requirement remains independent of speed, as shown by the dotted line in Figure 12.2b, the speed can be adjusted by controlling the frequency of the applied voltages; for example, the speed is $\omega_{m,1}(=\omega_{syn,1} - \omega_{slip,1})$ at a frequency of f_1, and $\omega_{m,2}(=\omega_{syn,2} - \omega_{slip,2})$ at f_2.

Example 12.1
A three-phase, 60-Hz, 4-pole, 440-V (line-line, rms) induction-motor drive has a full-load (rated) speed of 1746 rpm. The rated torque is 40 Nm. Keeping the air gap flux-density peak constant at its rated value, (a) plot the torque-speed characteristics (the linear portion) for the following values of the frequency f: 60 Hz, 45 Hz, 30 Hz, and 15 Hz. (b) This motor is supplying a load whose torque demand increases linearly with speed such that it equals the rated motor torque at the rated motor speed. Calculate the speeds of operation at the four values of frequency in part (a).

Solution
    a. In this example, it is easier to make use of speed (denoted by the symbol "n") in rpm. At the rated frequency of 60 Hz, the synchronous speed in a 4-pole motor can be calculated as follows: from Equation 12.5,

$$\omega_{syn,rated} = \frac{2\pi f_{rated}}{p/2}$$

Therefore,

$$n_{syn,rated} = \underbrace{\frac{\omega_{syn,rated}}{2\pi}}_{\text{rev. per sec.}} \times 60 \text{ rpm} = \frac{f_{rated}}{p/2} \times 60 \text{ rpm} = 1800 \text{ rpm}.$$

Therefore,

$$n_{slip,rated} = 1800 - 1746 = 54 \text{ rpm}.$$

The synchronous speeds corresponding to the other three frequency values are: 1350 rpm at 45 Hz, 900 rpm at 30 Hz, and 450 rpm at 15 Hz. The torque-speed characteristics are parallel, as shown in Figure 12.3, for the four frequency values, keeping $\hat{B}_{ms} = \hat{B}_{ms,rated}$.

b. The torque-speed characteristic in Figure 12.3 can be described for each frequency by the equation below, where n_{syn} is the synchronous speed corresponding to that frequency:

$$T_{em} = k_{Tn}(n_{syn} - n_m) \tag{12.7}$$

In this example, $k_{Tn} = \frac{40 \text{ Nm}}{(1800-1746)\text{rpm}} = 0.74 \frac{\text{Nm}}{\text{rpm}}$. The linear load torque-speed characteristic can be described as

$$T_L = c_n n_m \tag{12.8}$$

where, in this example, $c_n = \frac{40 \text{ Nm}}{1746 \text{ rpm}} = 0.023 \frac{\text{Nm}}{\text{rpm}}$.

In steady state, the electromagnetic torque developed by the motor equals the load toque. Therefore, equating the right sides of Equations 12.7 and 12.8,

$$k_{Tn}(n_{syn} - n_m) = c_n n_m. \tag{12.9}$$

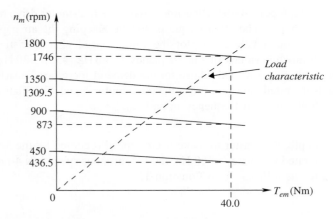

FIGURE 12.3   Example 12.1.

Hence,

$$n_m = \frac{k_{Tn}}{k_{Tn} + c_n} n_{syn} = 0.97 n_{syn} \text{ (in this example).} \qquad (12.10)$$

Therefore, we have the following speeds and slip speeds at various values of f:

f (Hz)	n_{syn} (rpm)	n_m (rpm)	n_{slip} (rpm)
60	1800	1746	54
45	1350	1309.5	40.5
30	900	873	27
15	450	436.5	13.5

12.3 APPLIED VOLTAGE AMPLITUDES TO KEEP $\hat{B}_{ms} = \hat{B}_{ms,rated}$

Maintaining $\hat{B}_{ms}$ at its rated value minimizes power loss in the rotor circuit. To maintain $\hat{B}_{ms,rated}$ at various frequencies and torque loading, the applied voltages should be of the appropriate amplitude, as discussed in this section.

The per-phase equivalent circuit of an induction motor under the balanced sinusoidal steady state is shown in Figure 12.4a. With the rated voltages at $\hat{V}_{a,rated}$ and f_{rated} applied to the stator, loading the motor by its rated (full-load) torque $T_{em,rated}$ establishes the rated operating point. At the rated operating point, all quantities related to the motor are at their rated values: the synchronous speed $\omega_{syn,rated}$, the motor speed $\omega_{m,rated}$, the slip

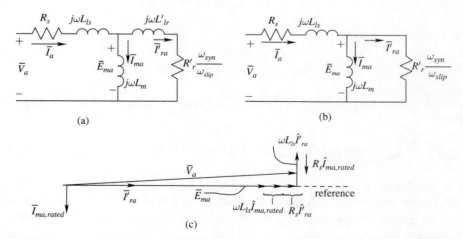

FIGURE 12.4 (a) Per-phase equivalent circuit in balanced steady state; (b) equivalent circuit with the rotor-leakage neglected; (c) phasor diagram in steady state at the rated flux-density.

speed $\omega_{slip,rated}$, the flux-density peak $\hat{B}_{ms,rated}$, the internal voltage $\hat{E}_{ma,rated}$, the magnetizing current $\hat{I}_{ma,rated}$, the rotor-branch current $\hat{I}'_{ra,rated}$, and the stator current $\hat{I}_{a,rated}$.

The objective of maintaining the flux density at $\hat{B}_{ms,rated}$ implies that in the equivalent circuit of Figure 12.4a, the magnetizing current should be maintained at $\hat{I}_{ma,rated}$:

$$\hat{I}_{ma} = \hat{I}_{ma,rated} \ (\text{a constant}) \tag{12.11}$$

With this magnetizing current, the internal voltage $\overline{E}_{ma}$ in Figure 12.4a has the following amplitude:

$$\hat{E}_{ma} = \omega L_m \hat{I}_{ma,rated} = \underbrace{2\pi L_m \hat{I}_{ma,rated}}_{\text{constant}} f \tag{12.12}$$

This shows that $\hat{E}_{ma}$ is linearly proportional to the frequency f of the applied voltages.

For torques below the rated value, the leakage inductance of the rotor can be neglected (see Example 11.6), as shown in the equivalent circuit of Figure 12.4b. With this assumption, the rotor-branch current $\hat{I}'_{ra}$ is in phase with the internal voltage $\overline{E}_{ma}$, and its amplitude $\hat{I}'_{ra}$ depends linearly on the electromagnetic torque developed by the motor (as in Equation 11.28) to provide the load torque. Therefore, in terms of the rated values,

$$\hat{I}'_{ra} = \left(\frac{T_{em}}{T_{em,rated}} \right) \hat{I}'_{ra,rated} \tag{12.13}$$

At some frequency and torque, the phasor diagram corresponding to the equivalent circuit in Figure 12.4b is shown in Figure 12.4c. If the internal emf is the reference phasor $\overline{E}_{ma} = \hat{E}_{ma} \angle 0°$, then $\hat{I}'_{ra} = \hat{I}'_{ra} \angle 0°$ and the applied voltage is

$$\overline{V}_a = \hat{E}_{ma} \angle 0° + (R_s + j2\pi f L_{\ell s})\overline{I}_s \tag{12.14}$$

where

$$\overline{I}_s = \hat{I}'_{ra} \angle 0° - j\hat{I}_{ma,rated} \tag{12.15}$$

Substituting Equation 12.15 into Equation 12.14 and separating the real and imaginary parts,

$$\overline{V}_a = [\hat{E}_{ma} + (2\pi f L_{\ell s})\hat{I}_{ma,rated} + R_s \hat{I}'_{ra}] + j[(2\pi f L_{\ell s})\hat{I}'_{ra} - R_s \hat{I}_{ma,rated}] \tag{12.16}$$

These phasors are plotted in Figure 12.4c near the rated operating condition, using reasonable parameter values. This phasor diagram shows that in determining the magnitude $\hat{V}_a$ of the applied voltage phasor $\overline{V}_a$, the perpendicular component in Equation 12.16 can be neglected, yielding

$$\hat{V}_a \cong \hat{E}_{ma} + (2\pi f L_{\ell s})\hat{I}_{ma,rated} + R_s \hat{I}'_{ra} \tag{12.17}$$

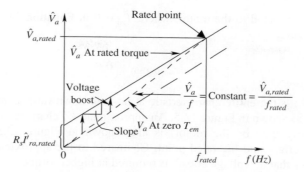

FIGURE 12.5 Relation of applied voltage and frequency at the rated flux density.

Substituting for $\hat{E}_{ma}$ from Equation 12.12 into Equation 12.17 and rearranging terms,

$$\hat{V}_a = \underbrace{2\pi(L_m + L_s)\,\hat{I}_{ma,rated}}_{\text{constant slope}} f + R_s\,\hat{I}'_{ra} \quad \text{or} \quad \hat{V}_a = (slope)f + R_s\,\hat{I}'_{ra} \tag{12.18}$$

This shows that to maintain flux density at its rated value, the applied voltage amplitude $\hat{V}_a$ depends linearly on the frequency f of the applied voltages, except for the offset due to the resistance R_s of the stator windings. At a constant torque value, the relationship in Equation 12.18 between $\hat{V}_a$ and f is a straight line, as shown in Figure 12.5.

This line has a constant slope equal to $2\pi(L_m + L_{\ell s})\,\hat{I}_{ma,rated}$. This slope can be obtained by using the values at the rated operating point of the motor in Equation 12.18:

$$slope = \frac{\hat{V}_{a,rated} - R_s\,\hat{I}'_{ra,rated}}{f_{rated}} \tag{12.19}$$

Therefore, in terms of the slope in Equation 12.19, the relationship in Equation 12.18 can be expressed as

$$\hat{V}_a = \left(\frac{\hat{V}_{a,rated} - R_s\,\hat{I}'_{ra,rated}}{f_{rated}}\right)f + R_s\,\hat{I}'_{ra} \tag{12.20}$$

At the rated torque, in Equation 12.20, $\hat{V}_a$, $\hat{I}'_{ra}$, and f are all at their rated values. This establishes the rated point in Figure 12.5. Continuing to provide the rated torque, as the frequency f is reduced to nearly zero at very low speeds, from Equation 12.20,

$$\hat{V}_a\big|_{T_{em,rated},f\simeq 0} = R_s\,\hat{I}'_{ra,rated} \tag{12.21}$$

This is shown by the offset above the origin in Figure 12.5. Between this offset point (at $f \simeq 0$) and the rated point, the voltage-frequency characteristic is linear, as shown, while the motor is loaded to deliver its rated torque. We will consider another case of no-load connected to the motor, where $\hat{I}'_{ra} \simeq 0$ in Equation 12.20, and hence at nearly zero frequency

$$\hat{V}_a|_{T_{em}=0,\,f\simeq0} = 0 \tag{12.22}$$

This condition shifts the entire characteristic at no-load downwards compared to that at the rated torque, as shown in Figure 12.5. An approximate V/f characteristic (independent of the torque developed by the motor) is also shown in Figure 12.5 by the dotted line through the origin and the rated point. Compared to the approximate relationship, Figure 12.5 shows that a "voltage boost" is required at higher torques, due to the voltage drop across the stator resistance. In percentage terms, this voltage boost is very significant at low frequencies, which correspond to operating the motor at low speeds; the percentage voltage boost that is necessary near the rated frequency (near the rated speed) is much smaller.

Example 12.2
In the motor drive of Example 12.1, the induction motor is such that while applied the rated voltages and loaded to the rated torque, it draws 10.39 A (rms) per-phase at a power factor of 0.866 (lagging). $R_s = 1.5\ \Omega$. Calculate the voltages corresponding to the four values of the frequency f to maintain $\hat{B}_{ms} = \hat{B}_{ms,rated}$.

Solution Neglecting the rotor leakage inductance, as shown in the phasor diagram of Figure 12.6, the rated value of the rotor-branch current can be calculated as

$$\hat{I}'_{ra,rated} = 10.39\sqrt{2}(0.866) = 9.0\sqrt{2}\ \text{A}.$$

Using Equation 12.20 and the rated values, the slope of the characteristic can be calculated as

$$slope = \frac{\hat{V}_{a,rated} - R_s \hat{I}'_{ra,rated}}{f_{rated}} = \frac{\frac{440\sqrt{2}}{\sqrt{3}} - 1.5 \times 9.0\sqrt{2}}{60} = 5.67\ \frac{\text{V}}{\text{Hz}}.$$

In Equation 12.20, $\hat{I}'_{ra}$ depends on the torque that the motor is supplying. Therefore, substituting for $\hat{I}'_{ra}$ from Equation 12.13 into Equation 12.20,

$$\hat{V}_a = \left(\frac{\hat{V}_{a,rated} - R_s \hat{I}'_{ra,rated}}{f_{rated}}\right)f + R_s \left(\frac{T_{em}}{T_{em,rated}}\, \hat{I}'_{ra,rated}\right). \tag{12.23}$$

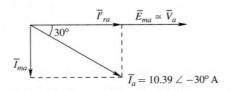

FIGURE 12.6 Example 12.2.

While the drive is supplying a load whose torque depends linearly on speed (and demands the rated torque at the rated speed as in Example 12.1), the torque ratio in Equation 12.23 is

$$\frac{T_{em}}{T_{em,rated}} = \frac{n_m}{n_{m,rated}}.$$

Therefore, Equation 12.23 can be written as

$$\hat{V}_a = \left(\frac{\hat{V}_{a,rated} - R_s \hat{I}'_{ra,rated}}{f_{rated}}\right)f + R_s\left(\frac{n_m}{n_{m,rated}}\hat{I}'_{ra,rated}\right) \quad (12.24)$$

Substituting the four values of the frequency f and their corresponding speeds from Example 12.1, the voltages can be tabulated as below. The values obtained by using the approximate dotted characteristic plotted in Figure 12.5 (which assumes a linear V/f relationship) are nearly identical to the values in the table below because at low values of frequency (hence, at low speeds) the torque is also reduced in this example—therefore, no voltage boost is necessary.

f	60 Hz	45 Hz	30 Hz	15 Hz
$\hat{V}_a$	359.3 V	269.5 V	179.6 V	89.8 V

12.4   STARTING CONSIDERATIONS IN DRIVES

Starting currents are primarily limited by the leakage inductances of the stator and the rotor, and can be 6 to 8 times the rated current of the motor, as shown in the plot of Figure 11.23b in Chapter 11. In the motor drives of Figure 12.1, if large currents are drawn even for a short time, the current rating required of the power-processing unit will become unacceptably large.

At starting, the rotor speed ω_m is zero, and hence the slip speed ω_{slip} equals the synchronous speed ω_{syn}. Therefore, at start-up, we must apply voltages of a low frequency in order to keep ω_{slip} low, and hence avoid large starting currents. Figure 12.7a shows the torque-speed characteristic at a frequency $f_{start}(=f_{slip,rated})$ of the applied voltages, such that the starting torque (at $\omega_m = 0$) is equal to the rated value. The same is true of the rotor-branch current. It is assumed that the applied voltage magnitudes are appropriately adjusted to maintain $\hat{B}_{ms}$ constant at its rated value.

As shown in Figure 12.7b, as the rotor speed builds up, the frequency f of the applied voltages is increased continuously at a preset rate until the final desired speed is reached in steady state. The rate at which the frequency is increased should not let the motor current exceed a specific limit (usually 150 percent of the rated). The rate should be decreased for higher inertia loads to allow the rotor speed to catch up. Note that the voltage amplitude is adjusted, as a function of the frequency f, as discussed in the previous section, to keep $\hat{B}_{ms}$ constant at its rated value.

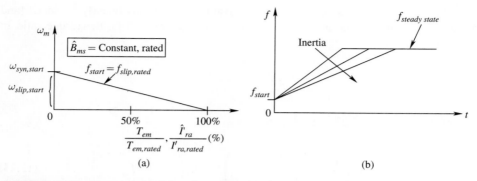

FIGURE 12.7   Start-up considerations in induction-motor drives.

Example 12.3

The motor drive in Examples 12.1 and 12.2 needs to develop a starting torque of 150 percent of the rated in order to overcome the starting friction. Calculate f_{start} and $\hat{V}_{a,start}$.

Solution   The rated slip of this motor is 54 rpm. To develop 150 percent of the rated torque, the slip speed at start-up should be $1.5 \times n_{slip,rated} = 81$ rpm. Note that at start-up, the synchronous speed is the same as the slip speed. Therefore, $n_{syn,start} = 81$ rpm. Hence, from Equation 12.5 for this 4-pole motor,

$$f_{start} = \underbrace{\left(\frac{n_{syn,start}}{60}\right)}_{\text{rev. per second}} \frac{p}{2} = 2.7 \text{ Hz}$$

At 150 percent of the rated torque, from Equation 12.13,

$$\hat{I}'_{ra,start} = 1.5 \times \hat{I}'_{ra,rated} = 1.5 \times 9.0\sqrt{2} \text{ A}.$$

Substituting various values at start-up into Equation 12.20,

$$\hat{V}_{a,start} = 43.9 \text{ V}$$

12.5   CAPABILITY TO OPERATE BELOW AND ABOVE THE RATED SPEED

Due to the rugged construction of the squirrel-cage rotor, induction-motor drives can be operated at speeds in the range of zero to almost twice the rated speed. The following constraints on the drive operation should be noted:

- The magnitude of applied voltages is limited to their rated value. Otherwise, the motor insulation may be stressed, and the rating of the power-processing unit will have to be larger.
- The motor currents are also limited to their rated values. This is because the rotor-branch current $\hat{I}'_{ra}$ is limited to its rated value in order to limit the loss $P_{r,loss}$ in the rotor

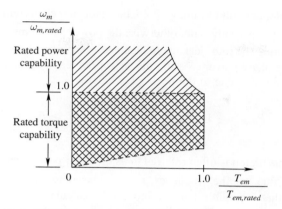

FIGURE 12.8 Capability below and above the rated speed.

bar resistances. This loss, dissipated as heat, is difficult to remove; beyond its rated value, it will cause the motor temperature to exceed its design limit, thus shortening the motor life.

The torque-capability regions below and above the rated speed are shown in Figure 12.8 and discussed in the following sections.

12.5.1 Rated Torque Capability below the Rated Speed (with $\hat{B}_{ms,rated}$)

This region of operation has already been discussed in Section 12.3 where the motor is operated at the rated flux density $\hat{B}_{ms,rated}$. Therefore at any speed below the rated speed, a motor in steady state can deliver its rated torque while $\hat{I}'_{ra}$ stays equal to its rated value. This capability region is shown in Figure 12.8 as the rated-torque capability region. At low speeds, due to poor cooling, the steady state torque capability may have to be reduced, as shown by the dotted curve.

12.5.2 Rated Power Capability above the Rated Speed by Flux-Weakening

Speeds above the rated value are obtained by increasing the frequency f of the applied voltages above the rated frequency, thus increasing the synchronous speed at which the flux-density distribution rotates in the air gap:

$$\omega_{syn} > \omega_{syn,rated} \tag{12.25}$$

The amplitude of the applied voltages is limited to its rated value $\hat{V}_{a,rated}$, as discussed earlier. Neglecting the voltage drop across the stator winding leakage inductance and resistance, in terms of the rated values, the peak flux density $\hat{B}_{ms}$ declines below its rated value, such that it is inversely proportional to the increasing frequency f (in accordance with Equation 11.7 of the previous chapter):

$$\hat{B}_{ms} = \hat{B}_{ms,rated}\frac{f_{rated}}{f} \quad (f > f_{rated}) \tag{12.26}$$

In the equivalent circuit of Figure 12.4b, the rotor-branch current should not exceed its rated value $\hat{I}'_{ra,rated}$ in steady state; otherwise the power loss in the rotor will exceed its rated value. Neglecting the rotor leakage inductance when estimating the capability limit, the maximum three-phase power crossing the air gap, in terms of the peak quantities (the additional factor of $1/2$ is due to the peak quantities) is

$$P_{max} = \frac{3}{2} \hat{V}_{a,rated} \hat{I}'_{a,rated} = P_{rated} \quad (f > f_{rated}) \tag{12.27}$$

Therefore, this region is often referred to as the rated-power capability region. With $\hat{I}'_{ra}$ at its rated value, as the frequency f is increased to obtain higher speeds, the maximum torque that the motor can develop can be calculated by substituting the flux density given by Equation 12.26 into Equation 11.28 of the previous chapter:

$$T_{em}\big|_{\hat{I}'_{r,rated}} = k_t \hat{I}'_{r,rated} \hat{B}_{ms} = \underbrace{k_t \hat{I}'_{r,rated} \hat{B}_{ms,rated}}_{T_{em,rated}} \frac{f}{f_{rated}} = T_{em,rated} \frac{f_{rated}}{f} \tag{12.28}$$

This shows that the maximum torque, plotted in Figure 12.8, is inversely proportional to the frequency.

12.6   INDUCTION-GENERATOR DRIVES

Induction machines can operate as generators, as discussed in Section 11.3.2.4. For an induction machine to operate as a generator, the applied voltages must be at a frequency at which the synchronous speed is less than the rotor speed, resulting in a negative slip speed:

$$\omega_{slip} = (\omega_{syn} - \omega_m) < 0 \qquad \omega_{syn} < \omega_m \tag{12.29}$$

Maintaining the flux density at $\hat{B}_{ms,rated}$ by controlling the voltage amplitudes, the torque developed, according to Equation 12.4, is negative (in a direction opposite that of rotation) for negative values of slip speed. Figure 12.9 shows the motor torque-speed characteristics at two frequencies, assuming a constant $\hat{B}_{ms} = \hat{B}_{ms,rated}$. These characteristics are extended into the negative torque region (generator region) for the rotor speeds above the corresponding synchronous speeds. Consider that the induction machine is initially operating as a motor with a stator frequency f_0 and at the rotor speed of ω_{m_0}, which is less than ω_{syn_0}. If the stator frequency is decreased to f_1, the new synchronous speed is ω_{syn_1}. This makes the slip speed negative, and thus T_{em} becomes negative, as shown in Figure 12.9. This torque acts in the direction opposite to the direction of rotation.

Therefore, in the wind-turbine application such as that shown in Figure 11.28 of the previous chapter, if a squirrel-cage induction machine is to operate as a generator, then the synchronous-speed ω_{syn} (corresponding to the frequency of the applied voltages to the motor terminals by the power electronics interface), must be less than the rotor-speed ω_m, so that the slip-speed ω_{slip} is negative, and the machine operates as a generator.

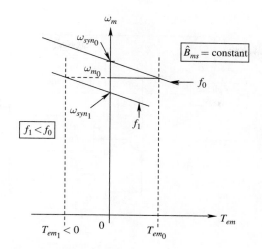

FIGURE 12.9 Induction-generator drives.

12.7 SPEED CONTROL OF INDUCTION-MOTOR DRIVES

The focus of this section is to discuss speed control of induction-motor drives in general-purpose applications where very precise speed control is not necessary, and therefore, as shown in Figure 12.10, the speed is not measured (rather it is estimated).

The reference speed $\omega_{m,ref}$ is set either manually or by a slow-acting control loop of the process where the drive is used. Use of induction-motor drives in high-performance servo-drive applications is discussed in the next chapter.

In addition to the reference speed, the other two inputs to the controller are the measured dc-link voltage V_d and the input current i_d of the inverter. This dc-link current represents the instantaneous three-phase currents of the motor. Some of the salient points of the control in Figure 12.10 are described below.

Limiting of Acceleration/Deceleration. During acceleration and deceleration, it is necessary to keep the motor currents and the dc-link voltage V_d within their design limits. Therefore, in Figure 12.10, the maximum acceleration and deceleration are usually set by the user, resulting in a dynamically-modified reference speed signal ω_m.

Current-Limiting. In the motoring mode, if ω_{syn} increases too fast compared to the motor speed, then ω_{slip} and the motor currents may exceed their limits. To limit acceleration so that the motor currents stay within their limits, i_d (representing the actual motor current) is compared with the current limit, and the error though the controller acts on the speed control circuit by reducing acceleration (i.e., by reducing ω_{syn}).

In the regenerative-braking mode, if ω_{syn} is reduced too fast, the negative slip will become too large in magnitude and will result in a large current through the motor and the inverter of the PPU. To restrict this current within the limit, i_d is compared with the current limit, and the error is fed through a controller to decrease deceleration (i.e., by increasing ω_{syn}).

During regenerative-braking, the dc-bus capacitor voltage must be kept within a maximum limit. If the rectifier of the PPU is unidirectional in power flow, a dissipation

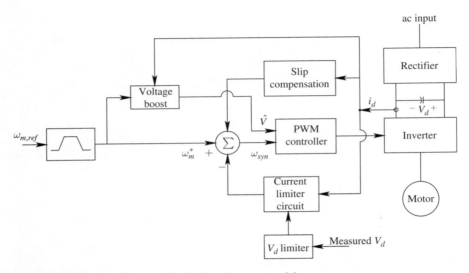

FIGURE 12.10   Speed control of induction motor drives.

resistor is switched on, in parallel with the dc-link capacitor, to provide a dynamic braking capability. If the energy recovered from the motor is still larger than that lost through various dissipation means, the capacitor voltage could become excessive. Therefore, if the voltage limit is exceeded, the control circuit decreases deceleration (by increasing ω_{syn}).

Slip Compensation. In Figure 12.10, to achieve a rotor speed equal to its reference value, the machine should be applied voltages at a frequency f, with a corresponding synchronous speed ω_{syn} such that it is the sum of ω_m^* and the slip speed:

$$\omega_{syn} = \omega_m^* + \underbrace{T_{em}/k_{T\omega}}_{\omega_{slip}} \qquad (12.30)$$

where the required slip speed, in accordance with Equation 12.4, depends on the torque to be developed. The slip speed is calculated by the slip-compensation block of Figure 12.10. Here, T_{em} is estimated as follows: the dc power input to the inverter is measured as a product of V_d and the average of i_d. From this, the estimated losses in the inverter of the PPU and in the stator resistance are subtracted to estimate the total power P_{ag} crossing the air gap into the rotor. We can show, by adding Equations 11.40 and 11.42 of the previous chapter that $T_{em} = P_{ag}/\omega_{syn}$.

Voltage Boost. To keep the air gap flux density $\hat{B}_{ms}$ constant at its rated value, the motor voltage must be controlled in accordance with Equation 12.18, where $\hat{I}_{ra}'$ is linearly proportional to T_{em} estimated earlier.

12.8   PULSE-WIDTH-MODULATED POWER-PROCESSING UNIT

In the block diagram of Figure 12.10, the inputs $\hat{V}$ and ω_{syn} generate the three control voltages that are compared with a switching-frequency triangular waveform v_{tri} of a

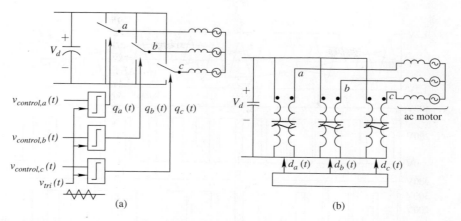

FIGURE 12.11 Power Processing Unit (a) switching representation; (b) average representation.

constant amplitude. The power-processing unit of Figure 12.11a, as described in Chapter 4, supplies the desired voltages to the stator windings. By averaging, each pole is represented by an ideal transformer in Figure 12.11b whose turns-ratio is continuously controlled, proportional to the control voltage.

12.8.1 Harmonics in the PPU Output Voltages

The instantaneous voltage waveforms corresponding to the logic signals are shown in Figure 12.12a. These are best discussed by means of computer simulations. The harmonic spectrum of the line-line output voltage waveform shows the presence of harmonic voltages as the sidebands of the switching frequency f_s and its multiples. The PPU output voltages, for example $v_a(t)$, can be decomposed into the fundamental-frequency component (designated by the subscript "1") and the ripple voltage

$$v_a(t) = v_{a1}(t) + v_{a,ripple}(t) \tag{12.31}$$

where the ripple voltage consists of the components in the range of and higher than the switching frequency f_s, as shown in Figure 12.12b. With the availability of higher switching-speed power devices such as modern IGBTs, the switching frequency in low and medium-power motor drives approach, and in some cases exceed, 20 kHz. The motivation for selecting a high switching frequency f_s, if the switching losses in the PPU can be kept manageable, is to reduce the ripple in the motor currents, thus reducing the electromagnetic torque ripple and the power losses in the motor resistances.

To analyze the motor's response to the applied voltages with ripple, we will make use of superposition. The motor's dominant response is determined by the fundamental-frequency voltages, which establish the synchronous speed ω_{syn} and the rotor speed ω_m. The per-phase equivalent circuit at the fundamental frequency is shown in Figure 12.13a.

In the PPU output voltages, the voltage components at a harmonic frequency $f_h \gg f$ produce rotating flux distribution in the air gap at a synchronous speed $\omega_{syn,h}$ where

$$\omega_{syn,h}(= h \times \omega_{syn}) \gg \omega_{syn}, \omega_m \tag{12.32}$$

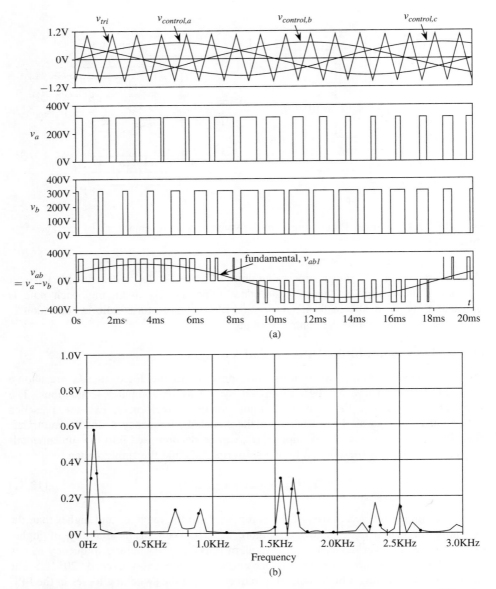

FIGURE 12.12 (a) PPU output voltage waveforms; (b) harmonic spectrum of line-to-line voltages.

The flux-density distribution at a harmonic frequency may be rotating in the same or opposite direction as the rotor. In any case, because it is rotating at a much faster speed compared to the rotor speed ω_m, the slip speed for the harmonic frequencies is

$$\omega_{slip,h} = \omega_{syn,h} \pm \omega_m \cong \omega_{syn,h} \tag{12.33}$$

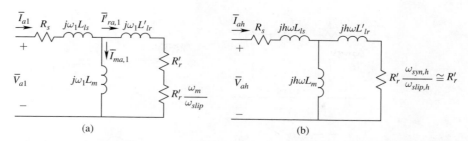

FIGURE 12.13 Per-phase equivalent circuit (a) at the fundamental frequency; (b) at harmonic frequencies.

Therefore, in the per-phase equivalent circuit at harmonic frequencies,

$$R'_r \frac{\omega_{syn,h}}{\omega_{slip,h}} \cong R'_r \qquad (12.34)$$

which is shown in Figure 12.13b. At high switching frequencies, the magnetizing reactance is very large and can be neglected in the circuit of Figure 12.13b, and the harmonic frequency current is determined primarily by the leakage reactances (which dominate over R'_r):

$$\hat{I}_{ah} \cong \frac{\hat{V}_{ah}}{X_{\ell s,h} + X'_{\ell r,h}} \qquad (12.35)$$

The additional power loss due to these harmonic frequency currents in the stator and the rotor resistances, on a three-phase basis, can be expressed as

$$\Delta P_{loss,R} = 3 \sum_h \frac{1}{2} (R_s + R'_r) \hat{I}_{ah}^2 \qquad (12.36)$$

In addition to these losses, there are additional losses in the stator and the rotor iron due to eddy currents and hysteresis at harmonic frequencies. These are further discussed in Chapter 15 dealing with efficiencies in drives.

12.8.2 Modeling the PPU-Supplied Induction Motors in Steady State

In steady state, an induction motor supplied by voltages from the power-processing unit should be modeled such that it allows the fundamental-frequency currents in Figure 12.13a and the harmonic-frequency currents in Figure 12.13b to be superimposed. This can be done if the per-phase equivalent is drawn as shown in Figure 12.14a, where the voltage drop across the resistance $R'_r \frac{\omega_m}{\omega_{slip}}$ in Figure 12.13a at the fundamental frequency is represented by a fundamental-frequency voltage $R'_r \frac{\omega_m}{\omega_{slip}} i'_{ra,1}(t)$. All three phases are shown in Figure 12.14b.

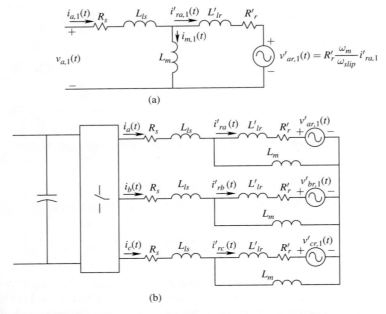

FIGURE 12.14 (a) Equivalent circuit for fundamental and harmonic frequencies in steady state; (b) three-phase equivalent circuit.

12.9   REDUCTION OF $\hat{B}_{ms}$ AT LIGHT LOADS

In Section 12.2, no attention was paid to core losses (only to the copper losses) in justifying that the machine should be operated at its rated flux density at any torque, while operating at speeds below the rated value. As illustrated by the discussion in Section 11.9 of Chapter 11, it is possible to improve the overall efficiency under lightly-loaded conditions by reducing $\hat{B}_{ms}$ below its rated value.

SUMMARY/REVIEW QUESTIONS

1. What are the applications of adjustable-speed drives?
2. Why are the thyristor-based, voltage reduction circuits for controlling induction-motor speed so inefficient?
3. In operating below the rated speed (and not considering the core losses), why is it most efficient to keep the flux-density peak in the air gap at the rated value?
4. Since an induction motor is operated at different values of frequency, hence different values of synchronous speed, how is the slip speed defined?
5. Supplying a load that demands a constant torque independent of speed, what is the slip speed at various values of the frequency f of the applied voltages?
6. To keep the flux-density peak in the air gap at the rated value, why do the voltage magnitudes, at a given frequency of operation, depend on the torque being supplied by the motor?

7. At start-up, why should small-frequency voltages be applied initially? What determines the rate at which the frequency can be ramped up?

8. At speeds below the rated value, what is the limit on the torque that can be delivered, and why?

9. At speeds above the rated value, what is the limit on the power that can be delivered, and why? What does it mean for the torque that can be delivered above the rated speed?

REFERENCES

1. N. Mohan, T. Undeland, and W. P. Robbins, *Power Electronics: Converters, Applications, and Design*, 2nd ed. (New York: John Wiley & Sons, 1995).

2. B. K. Bose, *Power Electronics and AC Drives* (Prentice-Hall, 1986).

3. M. Kazmierkowski, R. Krishnan and F. Blaabjerg, *Control of Power Electronics* (Academic Press, 2002).

PROBLEMS

12.1 Repeat Example 12.1 if the load is a centrifugal load that demands a torque, proportional to the speed squared, such that it equals the rated torque of the motor at the motor rated speed.

12.2 Repeat Example 12.2 if the load is a centrifugal load that demands a torque, proportional to the speed squared, such that it equals the rated torque of the motor at the motor rated speed.

12.3 Repeat Example 12.3 if the starting torque is to be equal to the rated torque.

12.4 Consider the drive in Examples 12.1 and 12.2, operating at the rated frequency of 60 Hz and supplying the rated torque. At the rated operating speed, calculate the voltages (in frequency and amplitude) needed to produce a regenerative braking torque that equals the rated torque in magnitude.

12.5 A 6-pole, 3-phase induction machine used for wind turbines has the following specifications: $V_{LL} = 600 \text{ V(rms)}$ at 60 Hz, the rated output power $P_{out} = 1.5 \text{ MW}$, the rated slip $s_{rated} = 1\%$. Assuming the machine efficiency to be approximately 95% while operating close to the rated power, calculate the frequency and the amplitude of the voltages to be applied to this machine by the power-electronics converter if the rotational speed is 1,100 rpm. Estimate the power output from this generator.

SIMULATION PROBLEM

12.6 Using the average representation of the PWM inverter, simulate the drive in Examples 12.1 and 12.2, while operating in steady state, at a frequency of 60 Hz. The dc bus voltage is 800 V, and the stator and the rotor leakage inductances are 2.2 Ω each. Estimate the rotor resistance R'_r from the data given in Examples 12.1 and 12.2.

13

RELUCTANCE DRIVES: STEPPER-MOTOR AND SWITCHED-RELUCTANCE DRIVES

13.1 INTRODUCTION

Reluctance machines operate on principles that are different from those associated with all of the machines discussed so far. Reluctance drives can be broadly classified into three categories: stepper-motor drives, switched-reluctance drives, and synchronous-reluctance-motor drives. Only the stepper-motor and the switched-reluctance-motor drives are discussed in this chapter.

Stepper-motor drives are widely used for position control in many applications, for example computer peripherals, textile mills, integrated-circuit fabrication processes, and robotics. A stepper-motor drive can be considered as a digital electromechanical device, where each electrical pulse input results in a movement of the rotor by a discrete angle called the step-angle of the motor, as shown in Figure 13.1. Therefore, for the desired change in position, the corresponding number of electrical pulses is applied to the motor, without the need for any position feedback.

Switched-reluctance-motor drives are operated with controlled currents, using feedback. They are being considered for a large number of applications discussed later in this chapter.

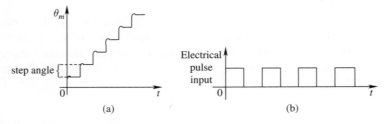

(a)   (b)

FIGURE 13.1 Position change in stepper motor.

13.2 THE OPERATING PRINCIPLE OF RELUCTANCE MOTORS

Reluctance motors operate by generating reluctance torque. This requires that the reluctance in the magnetic-flux path be different along the various axes. Consider the cross-section of a primitive machine shown in Figure 13.2a in which the rotor has no electrical excitation and the stator has a coil excited by a current $i(t)$. In the following analysis, we will neglect the losses in the electrical and the mechanical systems, but these losses can also be accounted for. In the machine of Figure 13.2a, the stator current would produce a torque on the rotor in the counter-clockwise direction, due to fringing fluxes, in order to align the rotor with the stator pole. This torque can be estimated by the principle of energy conservation; this principle states that

Electrical Energy Input = Increase in Stored Energy + Mechanical Output (13.1)

Assuming that magnetic saturation is avoided, the stator coil has an inductance $L(\theta)$ which depends on the rotor position θ. Thus, the flux-linkage λ of the coil can be expressed as

$$\lambda = L(\theta)i \qquad (13.2)$$

The flux-linkage λ depends on the coil inductance as well as the coil current. At any time, the voltage e across the stator coil, from Faraday's Law, is

$$e = \frac{d\lambda}{dt} \qquad (13.3)$$

The polarity of the induced voltage is indicated in Figure 13.2a. Based on Equations 13.2 and 13.3, the voltage in the coil may be induced due to the time-rate of change of the current and/or the coil inductance. Using Equation 13.3, the energy supplied by the electrical source from a time t_1 (with a flux linkage of λ_1) to time t_2 (with a flux linkage of λ_2) is

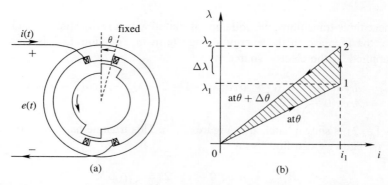

(a) (b)

FIGURE 13.2 (a) Cross-section of a primitive machine; (b) $\lambda - i$ trajectory during motion.

$$W_{el} = \int_{t_1}^{t_2} e \cdot i \cdot dt = \int_{t_1}^{t_2} \frac{d\lambda}{dt} \cdot i \cdot dt = \int_{\lambda_1}^{\lambda_2} i \cdot d\lambda \qquad (13.4)$$

In order to calculate the torque developed by this motor, we will consider the counter-clockwise movement of the rotor in Figure 13.2a by a differential angle $d\theta$ in the following steps shown in Figure 13.2b:

- Keeping θ constant, the current is increased from zero to a value i_1. The current follows the trajectory from 0 to 1 in the $\lambda - i$ plane in Figure 13.2b. Using Equation 13.4, we find that the energy supplied by the electrical source is obtained by integrating with respect to λ in Figure 13.2b; thus the energy supplied equals $Area\,(0 - 1 - \lambda_1)$

$$W_{el}(0 \rightarrow 1) = Area(0 - 1 - \lambda_1) = \frac{1}{2}\lambda_1 i_1 \qquad (13.5)$$

This is energy that gets stored in the magnetic field of the coil, since there is no mechanical output.
- Keeping the current constant at i_1, the rotor angle is allowed to be increased by a differential angle, from θ to $(\theta + \Delta\theta)$ in the counter-clockwise direction. This follows the trajectory from 1 to 2 in the $\lambda - i$ plane of Figure 13.2b. The change in flux linkage of the coil is due to the increased inductance. From Equation 13.2,

$$\Delta\lambda = i_1 \Delta L \qquad (13.6)$$

Using Equation 13.4 and integrating with respect to λ, we find that the energy supplied by the electrical source during this transition in Figure 13.2b is

$$W_{el}(1 \rightarrow 2) = Area(\lambda_1 - 1 - 2 - \lambda_2) = i_1(\lambda_2 - \lambda_1) \qquad (13.7)$$

- Keeping the rotor angle constant at $(\theta + \Delta\theta)$, the current is decreased from i_1 to zero. This follows the trajectory from 2 to 0 in the $\lambda - i$ plane in Figure 13.2b. Using Equation 13.4, we see that the energy is now supplied to the electrical source. Therefore, in Figure 13.2b,

$$W_{el}(2 \rightarrow 0) = -Area(2 - 0 - \lambda_2) = -\frac{1}{2}\lambda_2 i_1 \qquad (13.8)$$

During these three transitions, the coil current started with a zero value and ended at zero. Therefore, the increase in the energy storage term in Equation 13.1 is zero. The net energy supplied by the electric source is

$$\begin{aligned} W_{el,net} &= Area(0 - 1 - \lambda_1) + Area(\lambda_1 - 1 - 2 - \lambda_2) - Area(2 - 0 - \lambda_2) \\ &= Area(0 - 1 - 2) \end{aligned} \qquad (13.9)$$

$Area(0 - 1 - 2)$ is shown hatched in Figure 13.2b. This triangle has a base of $\Delta\lambda$ and a height of i_1. Thus we can find its area:

$$W_{el,net} = Area(0 - 1 - 2) = \frac{1}{2}i_1(\Delta\lambda) \qquad (13.10)$$

Using Equation 13.6 and Equation 13.10,

$$W_{el,net} = \frac{1}{2} i_1 (\Delta\lambda) = \frac{1}{2} i_1 (i_1 \cdot \Delta L) = \frac{1}{2} i_1^2 \Delta L \qquad (13.11)$$

Since there is no change in the energy stored, the electrical energy has been converted into mechanical work by the rotor, which is rotated by a differential angle $\Delta\theta$ due to the developed torque T_{em}. Therefore,

$$T_{em}\Delta\theta = \frac{1}{2} i_1^2 \Delta L \quad \text{or} \quad T_{em} = \frac{1}{2} i_1^2 \frac{\Delta L}{\Delta\theta} \qquad (13.12)$$

Assuming a differential angle,

$$T_{em} = \frac{1}{2} i_1^2 \frac{dL}{d\theta} \qquad (13.13)$$

This shows that the electromagnetic torque in such a reluctance motor depends on the current squared. Therefore, the counter-clockwise torque in the structure of Figure 13.2a is independent of the direction of the current. This torque, called the reluctance torque, forms the basis of operation for stepper motors and switched-reluctance motors.

13.3 STEPPER-MOTOR DRIVES

Stepper motors come in a large variety of constructions, with three basic categories: variable-reluctance motors, permanent-magnet motors, and hybrid motors. Each of these is briefly discussed.

13.3.1 Variable-Reluctance Stepper Motors

Variable-reluctance stepper motors have double saliency; that is, both the stator and the rotor have different magnetic reluctances along various radial axes. The stator and the rotor also have a different number of poles. An example is shown in Figure 13.3, in which the stator has six poles and the rotor has four poles. Each phase winding in this three-phase machine is placed on the two diametrically opposite poles.

Exciting phase-a with a current i_a results in a torque that acts in a direction to minimize the magnetic reluctance to the flux produced by i_a. With no load connected to the rotor, this torque will cause the rotor to align at $\theta = 0°$, as shown in Figure 13.3a. This is the no-load equilibrium position. If the mechanical load causes a small deviation in θ, the motor will develop an opposing torque in accordance with Equation 13.13.

To turn the rotor in a clockwise direction, i_a is reduced to zero and phase-b is excited by i_b, resulting in the no-load equilibrium position shown in Figure 13.3b. The point z on the rotor moves by the step-angle of the motor. The next two transitions with i_c and back to i_a are shown in Figs. 13.3c and 13.3d. Following the movement of point z, we see that the rotor has moved by one rotor-pole-pitch for three changes in excitation $(i_a \rightarrow i_b, i_b \rightarrow i_c, \text{and } i_c \rightarrow i_a)$. The rotor-pole-pitch equals $(360°/N_r)$, where N_r equals the number of rotor poles. Therefore, in a q-phase motor, the step-angle of rotation for each change in excitation will be

$$\text{step-angle} = \frac{360°}{qN_r} \qquad (13.14)$$

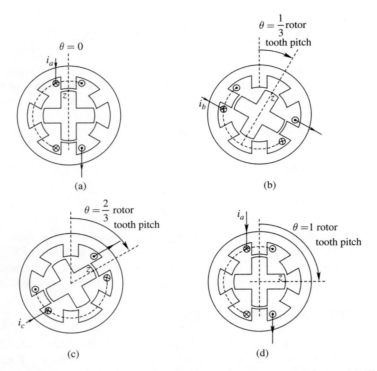

FIGURE 13.3 Variable-reluctance motor; excitation sequence a-b-c-a (a) Phase-a excited; (b) phase-b excited; (c) phase-c excited; (d) phase-a excited.

In the motor of Figure 13.3 with $N_r = 4$ and $q = 3$, the step-angle equals 30°. The direction of rotation can be made counter-clockwise by excitation in the sequence *a-c-b-a*.

13.3.2 Permanent-Magnet Stepper-Motors

In permanent-magnet stepper-motors, permanent magnets are placed on the rotor, as in the example shown in Figure 13.4. The stator has two phase windings. Each winding is placed on four poles, the same as the number of poles on the rotor. Each phase winding produces the same number of poles as the rotor. The phase currents are controlled to be positive or negative. With a positive current i_a^+, the resulting stator poles and the no-load equilibrium position of the rotor are as shown in Figure 13.4a. Reducing the current in phase-*a* to zero, a positive current i_b^+ in phase-*b* results in a clockwise rotation (following the point *z* on the rotor) shown in Figure 13.4b. To rotate further, the current in phase-*b* is reduced to zero, and a negative current i_a^- causes the rotor to be in the position shown in Figure 13.4c. Figure 13.4 illustrates that an excitation sequence $(i_a^+ \to i_b^+, i_b^+ \to i_a^-, i_a^- \to i_b^-, i_b^- \to i_a^+)$ produces a clockwise rotation. Each change in excitation causes rotation by one-half of the rotor-pole-pitch, which yields a step-angle of 45° in this example.

13.3.3 Hybrid Stepper-Motors

Hybrid stepper-motors utilize the principles of both the variable-reluctance and the permanent-magnet stepper-motors. An axial cross-section is shown in Figure 13.5.

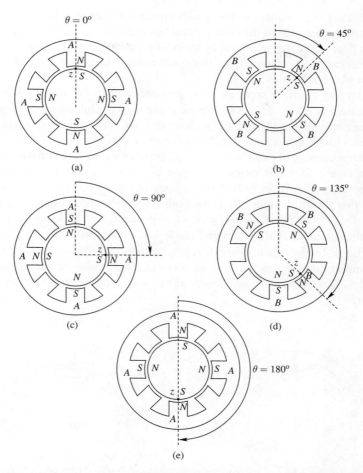

FIGURE 13.4 Two-phase permanent-magnet step-motor; excitation sequence $i_{a+}, i_{b+}, i_{a-}, i_{b-}, i_{a+}$ (a) i_{a+}; (b) i_{b+}; (c) i_{a-}; (d) i_{b-}; (e) i_{a+}.

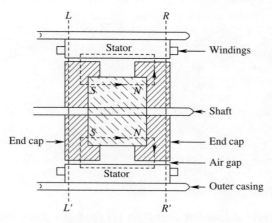

FIGURE 13.5 Axial view of a hybrid step-motor.

The rotor consists of permanent magnets with a north and a south pole at the two opposite ends. In addition, each side of the rotor is fitted with an end cap with N_r teeth; N_r is equal to 10 in this figure. The flux produced by the permanent magnets is shown in Figure 13.5. All of the end-cap teeth on the left act like south poles, while all of the end-cap teeth on the right act like north poles.

The left and the right cross-sections, perpendicular to the shaft, along $L-L'$ and $R-R'$, are shown in Figure 13.6. The two rotor end caps are intentionally displaced with respect to each other by one-half of the rotor-tooth-pitch. The stator in this figure consists of 8 poles in which the slots run parallel to the shaft axis.

The stator consists of two phases; each phase winding is placed on 4 alternate poles, as shown in Figure 13.6. Excitation of phase-a by a positive current i_a^+ results in north and south poles, as shown in both cross-sections in Figure 13.6a. In the no-load equilibrium position shown in Figure 13.6a, on both sides, the opposite stator and rotor poles align while the similar poles are as far apart as possible. For a clockwise rotation, the current in phase-a is brought to zero and phase-b is excited by a positive current i_b^+, as shown in Figure 13.6b. Again, on both sides, the opposite stator and rotor poles align while the similar poles are as far apart as possible. This change of excitation $(i_a^+ \rightarrow i_b^+)$ results in clockwise rotation by one-fourth of the rotor-tooth-pitch. Therefore, in a two-phase motor,

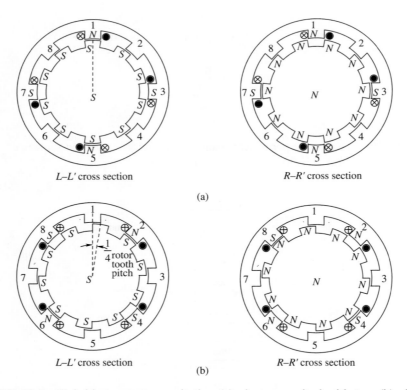

FIGURE 13.6 Hybrid step-motor excitation (a) phase-a excited with i_{a+}; (b) phase-b excited with i_{b+}

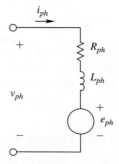

FIGURE 13.7 Per-phase equivalent circuit of a step-motor.

$$\text{step-angle} = \frac{360°/N_r}{4} \tag{13.15}$$

which, in this example with $N_r = 10$, equals 9°.

13.3.4 Equivalent-Circuit Representation of a Stepper-Motor

Similar to other machines discussed previously, stepper-motors can be represented by an equivalent circuit on a per-phase basis. Such an equivalent circuit for phase-a is shown in Figure 13.7 and consists of a back-emf, a winding resistance R_s, and a winding inductance L_s. The magnitude of the induced emf depends on the speed of rotation, and the polarity of the induced emf is such that it absorbs power in the motoring mode.

13.3.5 Half-Stepping and Micro-Stepping

It is possible to get smaller angular movement for each transition in the stator currents. For example, consider the variable-reluctance motor for which the no-load equilibrium positions with i_a and i_b were shown in Figures 13.3a and 13.3b, respectively. Exciting phases a and b simultaneously causes the rotor to be in the position shown in Figure 13.8, which is one-half of a step-angle away from the position in Figure 13.3a with i_a. Therefore, if "half-stepping" in the clockwise direction is required in the motor of Figure 13.3, the excitation sequence will be as follows:

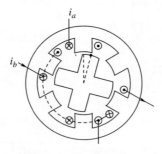

FIGURE 13.8 Half-excitation by exciting two phases.

$$i_a \rightarrow (i_a, i_b) \rightarrow i_b \rightarrow (i_b, i_c) \rightarrow i_c \rightarrow (i_c, i_a) \rightarrow i_a \qquad (13.16)$$

By precisely controlling the phase currents, it is possible to achieve micro-step angles. For example, there are hybrid-stepper motors in which a step-angle can be divided into 125 micro-steps. This results in 25,000 micro-steps/revolution in a two-phase hybrid motor with a step-angle of 1.8°.

13.3.6 Power-Processing Units for Stepper-Motors

In variable-reluctance drives, the phase currents need not reverse direction. A unidirectional-current converter for such motors is shown in Figure 13.9a. Turning both switches on simultaneously causes the phase current to build up quickly, as shown in Figure 13.9b. Once the current builds up to the desired level, it is maintained at that level by pulse-width-modulating one of the switches (for example T_1) while keeping the other switch on. By turning both switches off, the current is forced to flow into the dc-side source through the two diodes, thus decaying quickly.

   Bi-directional currents are needed in permanent-magnet and hybrid stepper-motors. Supplying these currents requires a converter such as that shown in Figure 13.10. This converter is very similar to those used in dc-motor drives discussed in Chapter 7.

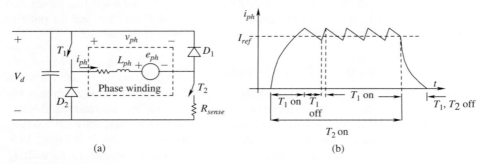

FIGURE 13.9 Unipolar voltage drive for variable-reluctance motor (a) Circuit; (b) Current waveform.

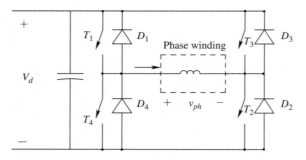

FIGURE 13.10 Bipolar voltage drive.

13.4 SWITCHED-RELUCTANCE MOTOR DRIVES

Switched-reluctance motors are essentially variable-reluctance stepper-motors that are operated in a closed-loop manner with controlled currents. In these drives, the appropriate phases are energized or de-energized based on the rotor position. These drives can potentially compete with other servo and adjustable-speed drives in a variety of applications.

Consider the cross-section shown in Figure 13.11. This motor is similar to the variable-reluctance stepper motor of Figure 13.3. At $t = 0$, the rotor is at an angle of $\theta = -\pi/6$ and the inductance of the phase-a winding is small due to a large air gap in the path of the flux lines. In order to move the rotor in Figure 13.11a counter-clockwise, the current i_a is built up quickly while the inductance is still small. As the rotor moves counter-clockwise, the phase-a inductance increases due to the stator and the rotor poles moving towards alignment, as shown in Figure 13.11b at $\theta = 0$. This increases the flux-linkage, causing the magnetic structure to go into a significant degree of saturation. Once at $\theta = 0$, phase-a is de-energized. The $\lambda - i$ trajectories for $\theta = -\pi/6$ and $\theta = 0$ are shown in Figure 13.11c. There are, of course, trajectories at the intermediate values of θ. The shaded area between the two trajectories in Figure 13.11c represents the energy that is converted into mechanical work. A sequence of excitations similar to that in the variable-reluctance drive of Figure 13.3 follows.

In order for switched-reluctance motors to be able to compete with other drives, they must be designed to go into magnetic saturation. A unidirectional-current converter such as that in Figure 13.9a can be used to power these motors.

There are many applications in which switched-reluctance drives may find their rightful place—from washing machines to automobiles to airplanes. Some of the strengths of switched-reluctance drives are their rugged and inexpensive rotor construction and their simple and reliable power electronics converter. On the negative side, these machines, due to their double saliency, produce large amounts of noise and vibrations.

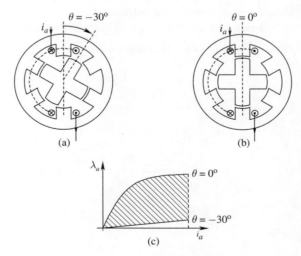

FIGURE 13.11   (a) Rotor at $\theta = -30°$; (b) Rotor at $\theta = 0°$; (c) $\lambda - i$ trajectory.

SUMMARY/REVIEW QUESTIONS

1. What are the three broad categories of reluctance drives?
2. How is the principle on which reluctance drives operate different from that seen earlier with other drives?
3. Write down the reluctance torque expression. What does the direction of torque depend on?
4. Describe the operating principle of a variable-reluctance stepper-motor.
5. Describe the operating principle of a permanent-magnet stepper-motor.
6. Describe the operating principle of a hybrid stepper-motor.
7. What is the equivalent-circuit representation of a stepper-motor?
8. How is half-stepping and micro-stepping achieved in stepper-motors?
9. What is the nature of power-processing units in stepper-motor drives?
10. Describe the operating principles of switched-reluctance drives.
11. What are the application areas of switched-reluctance drives?

REFERENCES

1. Takashi Kenjo, *Stepping Motors and Their Microprocessor Control* (Oxford: Oxford Science Publications, Clarendon Press, 1985).
2. P. P. Acarnley, *Stepping Motors: A guide to Modern Theory and Practice,* rev. 2nd ed. (IEE Control Engineering Series 19, 1984).
3. G. R. Slemon, *Electrical Machines for Drives*, Chapter 2, "Power Electronics and Variable Frequency Drives," edited by B. K. Bose (IEEE Press, 1997).

PROBLEMS

13.1   Determine the phase excitation sequence, and draw the rotor positions, in a variable-reluctance step drive for a counter-clockwise rotation.
13.2   Repeat Problem 13.1 for a permanent-magnet stepper-motor drive.
13.3   Repeat Problem 13.1 for a hybrid stepper-motor drive.
13.4   Describe the half-step operation in a permanent-magnet stepper-motor drive.
13.5   Describe the half-step operation in a hybrid stepper-motor drive.

14

ENERGY EFFICIENCY OF ELECTRIC DRIVES AND INVERTER-MOTOR INTERACTIONS

14.1 INTRODUCTION

Electric drives have enormous potential for improving energy efficiency in motor-driven systems. A market assessment of industrial electric-motor systems in the United States contains some startling, call-for-action statistics:

- Industrial motor systems consume 25 percent of the nation's electricity, making them the largest single electrical end-use.
- Potential yearly energy savings using mature, proven, and cost-effective technologies could equal the annual electricity use in the entire state of New York.

To achieve these energy savings would require a variety of means, but chief among them are the replacement of standard-efficiency motors by premium-efficiency motors and the use of variable-speed electric drives to improve system efficiencies.

The objective of this chapter is to briefly discuss the energy efficiencies of electric motors and electric drives over a range of loads and speeds. Since induction machines are the workhorses of industry, the discussion is limited to induction motors and induction-motor drives. The economics of investing in energy-efficient means are discussed. The interactions between induction motors and PWM inverters are briefly described as well.

The bulk of the material in this chapter is based on References [1] and [2]. A survey of the recent status and future trends is provided in [3].

14.2 THE DEFINITION OF ENERGY EFFICIENCY IN ELECTRIC DRIVES

As we briefly discussed in Chapter 6, the efficiency of an electric drive η_{drive} at an operating condition is the product of the corresponding motor efficiency η_{motor} and the PPU efficiency η_{PPU}:

$$\eta_{drive} = \eta_{motor} \times \eta_{PPU} \qquad (14.1)$$

In Equation 14.1, note that η_{motor} is the efficiency of a PPU-supplied motor. The output voltages of a power-processing unit consist of switching frequency harmonics, which usually lower the motor efficiency by one to two percentage points compared to the efficiency of the same motor when supplied by a purely sinusoidal source.

In the following section, we will look at the loss mechanisms and the energy efficiencies of induction motors and power-processing units.

14.3 THE ENERGY EFFICIENCY OF INDUCTION MOTORS WITH SINUSOIDAL EXCITATION

Initially, we will look at various loss mechanisms and energy efficiencies of motors with sinusoidal excitation, while later we will discuss the effects of switching-frequency harmonics of the PPU on motor losses.

14.3.1 Motor Losses

Motor power losses can be divided into four categories: core losses, winding losses, friction and windage losses, and stray-load losses. We will now briefly examine each of these.

14.3.1.1 Magnetic Core Losses

Magnetic losses are caused by hysteresis and eddy-currents in the magnetic core of the stator and the rotor. The losses depend on the frequency and the peak flux-density. Eddy-current losses can be reduced by thinner gauge steel laminations, between 0.014 and 0.025 inches thick, but at the expense of a higher assembly cost. Hysteresis losses cannot be reduced by thinner laminations but can be reduced by utilizing materials such as silicon steels with improved core-loss characteristics. For sinusoidal excitation at the rated slip, the loss in the rotor core is very small because the frequency of the flux variation in the rotor core is at the slip frequency and is very small. The magnetic core losses typically comprise 20 to 25 percent of the total motor losses at the rated voltage and frequency.

14.3.1.2 Winding Power Losses

These losses occur due to the ohmic (i^2R) heating of the stator winding and the rotor bars. The stator-winding loss is due to the sum of the magnetizing current and the torque-component of the stator current. This loss can be reduced by using larger cross-section conductors in the stator winding and by reducing the magnetizing current component. In the rotor, reducing the bar-resistances causes the motor to run closer to the synchronous speed, thus reducing the losses in the rotor bars. At full-load, the losses in the rotor bars are comparable to those in the stator winding, but drop to almost zero at no-load (although they do not do so in the presence of the switching-frequency harmonics of the PPU). At full-load, the combined stator and the rotor (i^2R) losses typically comprise 55 to 60 percent of the total motor losses.

14.3.1.3 Friction and Windage Losses

Losses in the bearings are caused by friction; windage losses are caused by the cooling fan and the rotor assembly. These losses are relatively fixed and can be reduced only indirectly by reducing the ventilation needed, which in turn is done by lowering other losses. These losses typically contribute 5 to 10 percent of the total motor losses.

14.3.1.4 Stray-Load Losses

This is a catch-all category for the losses that cannot be accounted for in any of the other three categories. These losses are load-dependent and vary as the square of the output torque. They typically contribute 10 to 15 percent of the total motor losses.

14.3.2 Dependence of Motor Losses and Efficiency on the Motor Load (with the Speed Essentially Constant)

A typical loss versus load curve is shown in Figure 14.1a. It shows that the core loss and the friction and windage losses are essentially independent of the load, whereas the stray-load losses and the winding losses vary as the square of the load.

A typical efficiency versus load plot is shown in Figure 14.1b. At a nominal voltage and frequency, most motors reach their maximum efficiency near the rated load. The efficiency remains nearly constant down to a 50 percent load and then falls rapidly down to zero below that level.

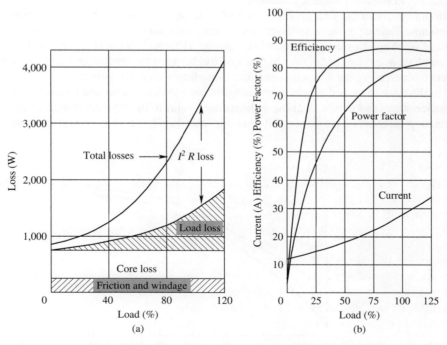

FIGURE 14.1   (a) Typical loss vs. load characteristics for Design B, 50-hp, 4-pole, 3-phase induction motor; (b) typical performance curves for Design B 10-hp, 4-pole, 3-phase induction motor.

14.3.3 Dependence of Motor Losses and Efficiency on Motor Speed (with the Torque Essentially Constant)

If an induction machine is operated from a variable-frequency sinusoidal source, the motor losses for constant-torque operation (assuming a constant air-gap flux) will vary as follows:

- Core losses are reduced at lower speeds because of the reduced frequencies.
- Stator-winding losses remain approximately unchanged because constant torque requires a constant current.
- Rotor-bar losses remain approximately unchanged because constant torque requires constant bar currents at a constant slip speed.
- Friction and windage losses are reduced at lower speeds.
- Stray-load losses are reduced at lower speeds.

We can see that the total losses drop as the frequency is reduced. Depending on whether the losses drop faster or slower than the output, the efficiency of the machine can increase or decrease with speed. The published literature in 40–400 horsepower machines indicates that for a constant torque, their efficiency is nearly constant down to 20 percent speed and exhibits a rapid drop toward zero below this speed level. With pump-loads requiring torque proportional to speed squared, the motor efficiency drops gradually to about 50 percent of speed and drops rapidly below that speed.

14.3.3.1 Premium-Efficiency Motors

With the advent of the Energy Policy Act of 1992, several manufactures have developed premium-efficiency motors. In these motors, the motor losses are reduced to typically 50 percent of those in the standard NEMA design B motors. This reduction in losses is accomplished by using thinner, higher-quality laminations, reducing the flux-density levels by increasing the core cross-section, using larger conductors in the stator windings and in the rotor cage, and carefully choosing the air gap dimensions and lamination design to reduce stray-load losses. Because of the reduced value of the rotor resistance, these high-efficiency machines have lower full-load slip speeds. Figure 14.2 shows a comparison

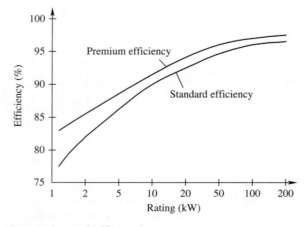

FIGURE 14.2 Comparison of efficiencies.

between the nominal efficiencies of standard-efficiency motors and premium-efficiency motors as a function of their power ratings. The typical increase in efficiency is 2 percentage points.

Typically, the power factor of operation associated with premium-efficiency motors is similar to that of motors of standard design; the power factor of premium-efficiency motors is slightly higher than that of standard motors at smaller power ratings and slightly lower at larger power ratings.

14.4 THE EFFECTS OF SWITCHING-FREQUENCY HARMONICS ON MOTOR LOSSES

All motor-loss components, except friction and windage, are increased as a result of the inverter-produced harmonics associated with the power-processing unit. For typical inverter waveforms, the total increase in losses is in the range of 10 to 20 percent and results in a decrease in energy efficiency of 1 to 2 percentage points at full load. Due to harmonics, increases in the various loss components are as follows:

- The core losses are slightly increased because of the slightly higher peak flux density caused by the superimposed harmonics. This increase is often negligibly small compared to other losses arising due to inverter harmonics.
- The stator-winding loss is increased due to the sum of the $(i^2 R)$ losses associated with the additional harmonic currents. At the harmonic frequencies, the stator resistance may be larger in bigger machines due to the skin effect. The increase in stator-winding loss is usually significant, but it is not the largest harmonic loss.
- The rotor-cage loss is increased due to the sum of $(i^2 R)$ losses associated with the additional harmonic currents. In large machines at harmonic frequencies, the deep-bar effect (similar to the skin effect) can greatly increase the rotor resistance and cause large rotor $(i^2 R)$ losses. These losses are often the largest loss attributable to harmonics.
- The stray-load losses are significantly increased by the presence of harmonic currents. These losses are the least understood, requiring considerable research activity.

In all cases, the harmonic losses are nearly independent of the load, because the harmonic slip is essentially unaffected by slight speed changes (in contrast to the fundamental slip).

In pulse-width-modulated inverters, the harmonic components of the output voltage depend on the modulation strategy. Further, the harmonic currents are limited by the machine-leakage inductances. Therefore, inverters with improved pulse-width-modulation strategies and machines with higher leakage inductances help to reduce these harmonic losses.

14.4.1 Motor De-Rating Due to Inverter Harmonic Losses

The increase in losses caused by inverter harmonics requires some de-rating of the motor in order to avoid overheating. It is often recommended that this "harmonic de-rating" be 10 percent of name-plate rating. Recently, many manufacturers have introduced inverter-grade motors that need not be de-rated.

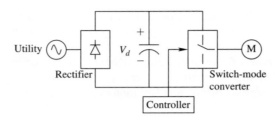

FIGURE 14.3   Block diagram of PPUs.

14.5   THE ENERGY EFFICIENCIES OF POWER-PROCESSING UNITS

The block diagram of a typical power-processing unit is shown in Figure 14.3. It consists of a diode-rectifier bridge to rectify line-frequency ac into dc and a switch-mode inverter to synthesize input dc into three-phase ac of adjustable magnitude and frequency.

Approximately 1 to 2 percent of the power is lost as conduction losses in the diode-rectifier bridge. The conduction and switching losses in the inverter total approximately 3 to 4 percent of the total power. Therefore, typical power loss in the PPU is in the range of 4 to 6 percent, resulting in the full-load PPU energy efficiency η_{PPU} in the range of 94 to 96 percent.

14.6   ENERGY EFFICIENCIES OF ELECTRIC DRIVES

Very little data is available to show the trend of the efficiencies of electric drives. A recent paper, however, shows that at full speed and full torque, the drive efficiency from a variety of manufacturers varies in the range of 74 to 80 percent for a 3-hp drive and in the range 86 to 89 percent for a 20-hp drive. At half-torque and half-speed (at one-fourth power), these efficiencies drop to 53 to 72 percent for a 3-hp drive and to 82 to 88 percent for a 20-hp drive. However, it is possible to modify the drive PPU so as to keep the energy efficiency high at light loads by slightly reducing the amplitude of the fundamental-frequency voltage.

14.7   THE ECONOMICS OF ENERGY SAVINGS BY PREMIUM-EFFICIENCY ELECTRIC MOTORS AND ELECTRIC DRIVES

In constant-speed applications, energy efficiency can be improved by replacing standard-design motors with premium-efficiency motors. In systems with dampers and throttling valves, and in compressors with on/off cycling, the use of adjustable-speed drives can result in dramatic savings in energy and thus savings in the cost of electricity. These savings accrue at the expense of the higher initial investment of replacing a standard motor, either with a slightly more expensive but more efficient motor, or with an adjustable-speed electric drive. Therefore, a user must consider the economics of initial investment—the payback period, at which the initial investment will have paid for itself, and the subsequent savings are money-in-the-bank.

14.7.1 The Present Worth of Savings and the Payback Period

The energy savings E_{save} take place every year over the period that the system is in operation. The present worth of these energy savings depends on many factors, such as the present cost of electricity, the rate of increase of the electricity cost, and the rate of investment of the money that could have been invested elsewhere. Inflation is another factor. Based on these factors, the present worth of the savings over the lifetime of the system can be obtained and compared to the additional initial investment. For a detailed discussion of this, Reference [5] is an excellent source. However, we can get an approximate idea of the payback period of the additional initial investment, if we ignore all of the previously mentioned factors and simply divide the additional initial investment by the yearly operational savings. This is illustrated by the following simple example.

Example 14.1

Calculate the payback period for investing in a premium-efficiency motor that costs $300 more than the standard motor, given the following parameters: the load demands a power of 25 kW, the efficiency of the standard motor is 89 percent, the efficiency of the premium-efficiency motor is 92 percent, the cost of electricity is 0.10 $/kWh, and the annual operating time of the motor is 4500 hours.

Solution

    a. The power drawn by the standard motor would be

$$P_{in} = \frac{P_0}{\eta_{motor}} = \frac{25.0}{0.89} = 28.09 \text{ kW}$$

    Therefore, the annual cost of electricity would be Annual Electricity Cost = 28.09 × 4500 × 0.1 = $12,640

    b. The power drawn by the premium-efficiency motor would be

$$P_{in} = \frac{P_0}{\eta_{motor}} = \frac{25.0}{0.92} = 27.17 \text{ kW}$$

    Therefore, the annual cost of electricity would be Annual Electricity Cost = 27.17 × 4500 × 0.1 = $12,226

    Thus, the annual savings in the operating cost = 12,640 − 12,226 = $414. Therefore, the initial investment of $300 would be paid back in

$$\frac{300}{414} \times 12 \approx 9 \text{ months}$$

14.8 THE DELETERIOUS EFFECTS OF THE PWM-INVERTER VOLTAGE WAVEFORM ON MOTOR LIFE

In supplying motors with PWM inverters, there are certain factors that users must be aware of. The additional harmonic losses due to the pulsating waveforms have already been discussed. If the motor is not specifically designed to operate with PWM inverters, it may be necessary to de-rate it (by a factor of 0.9) in order to accommodate addition harmonic losses without exceeding the motor's normal operating temperature.

The PWM-inverter output, particularly due to the ever-increasing switching speeds of IGBTs (which are good for keeping switching losses low in inverters), results in pulsating voltage waveforms with a very high dv/dt. These rapid changes in the output voltage have several deleterious effects: they stress the motor-winding insulation, they cause the flow of currents through the bearing (which can result in pitting), and they cause voltage doubling at the motor terminal due to the long cables between the inverter and the motor. One practical but limited solution is to attempt to slow down the switching of IGBTs at the expense of higher switching losses within the inverter. The other solution, which requires additional expense, is to add a small filter between the inverter and the motor.

14.9   BENEFITS OF USING VARIABLE-SPEED DRIVES

Readers are urged to look at Reference [6], an excellent source of information on the potential of achieving higher energy efficiencies using variable-speed drives and permanent-magnet machines in the residential sector.

SUMMARY/REVIEW QUESTIONS

1. What is the definition of energy efficiency of electric drives?
2. What are the various mechanisms of losses in motors, assuming a sinusoidal excitation?
3. How do the losses and the efficiency depend on motor speed, assuming a constant torque loading?
4. What are premium-efficiency motors? How much more efficient are they, compared to standard motors?
5. What are the effects of switching-frequency harmonics on the motor? How much should the motor be de-rated?
6. What is the typical range associated with the energy efficiency of power-processing units and of overall drives?
7. Discuss the economics and the payback period of using premium-efficiency motors.
8. Describe the various deleterious effects of PWM-inverter output voltage waveforms. Describe the techniques for mitigating these effects.

REFERENCES

1. N. Mohan, *Techniques for Energy Conservation in AC Motor-Driven Systems*, EPRI Final Report EM-2037, Project 1201–13, September 1981.
2. N. Mohan and J. Ramsey, *A Comparative Study of Adjustable-Speed Drives for Heat Pumps*, EPRI Final Report EM-4704, Project 2033–4, August 1986.
3. P. Thogersen Mohan and F. Blaabjerg, "Adjustable-Speed Drives in the Next Decade: The Next Steps in Industry and Academia," Proceedings of the PCIM Conference, Nuremberg, Germany, June 6–8, 2000.
4. G. R. Slemons, *Electrical Machines for Drives, in Power Electronics and Variable Frequency Drives*, edited by B. K. Bose (IEEE Press. 1997).

5. J. C. Andreas, *Energy-Efficient Electric Motors: Selection and Application* (New York: Marcel Dekker, 1982).

6. D. M. Ionel, "High-efficiency variable-speed electric motor drive technologies for energy savings in the US residential sector," 12th International Conference on Optimization of Electrical and Electronic Equipment, OPTIM 2010, Brasov, Romania, ISSN: 1842–0133.

PROBLEM

14.1   Repeat Example 14.1 if the motor runs fully loaded for one-half of each day and is shutoff for the other half.

INDEX